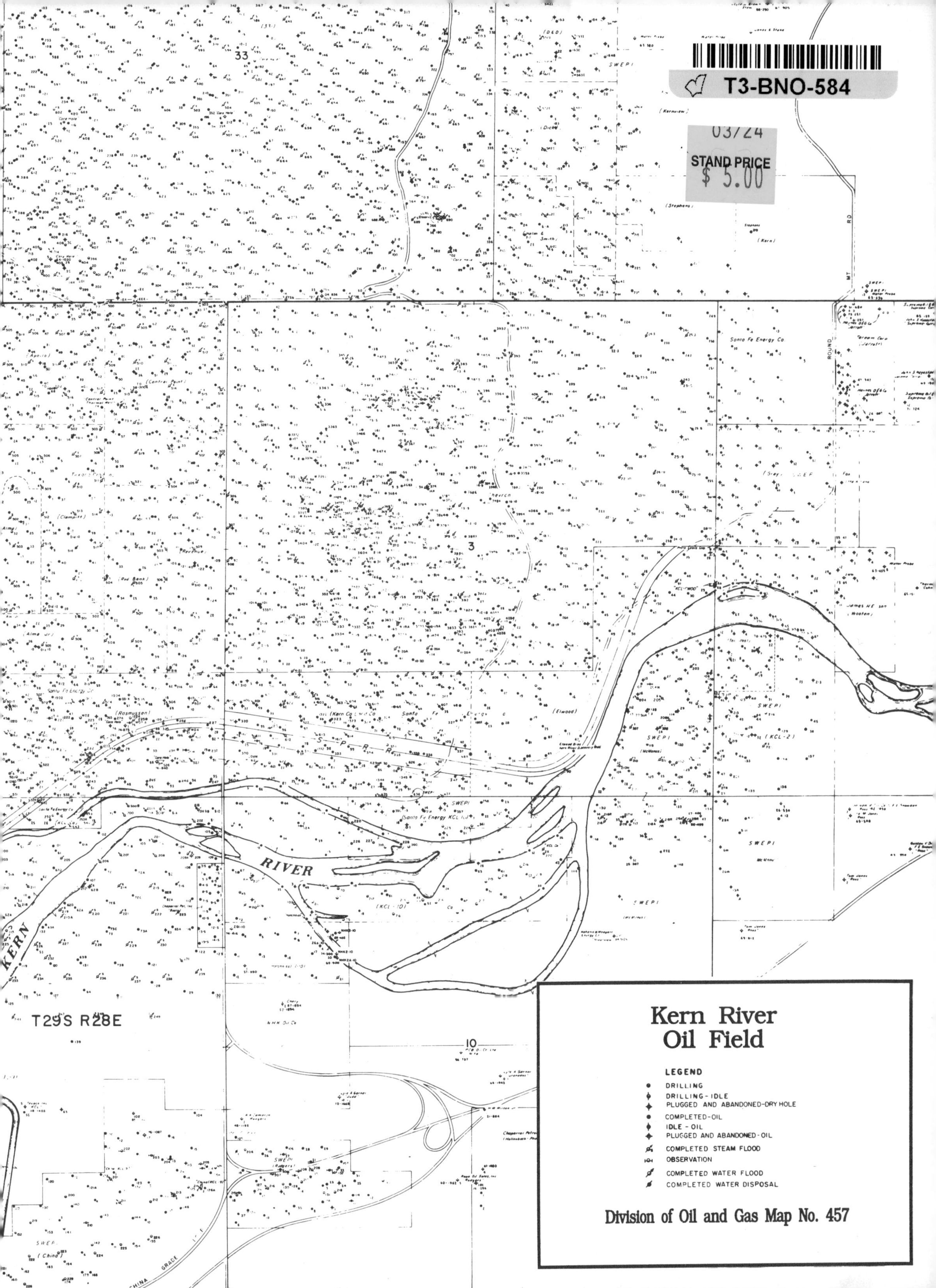
T3-BNO-584
03/24
STAND PRICE
$ 5.00
33
3
10
RIVER
KERN
S P R R
Santa Fe Energy Co
SWEPI
ROUND MT RD
CHINA GRADE
T29S R28E
Kern River
Oil Field
LEGEND
DRILLING
DRILLING-IDLE
PLUGGED AND ABANDONED-DRY HOLE
COMPLETED-OIL
IDLE-OIL
PLUGGED AND ABANDONED-OIL
COMPLETED STEAM FLOOD
OBSERVATION
COMPLETED WATER FLOOD
COMPLETED WATER DISPOSAL
Division of Oil and Gas Map No. 457

For Tom and Sue
Wishing you all the best always
Bill Rintoul
December 2, 1990

Alfred Zucca

DRILLING THROUGH TIME

75 years with California's Division of Oil and Gas

by

William Rintoul

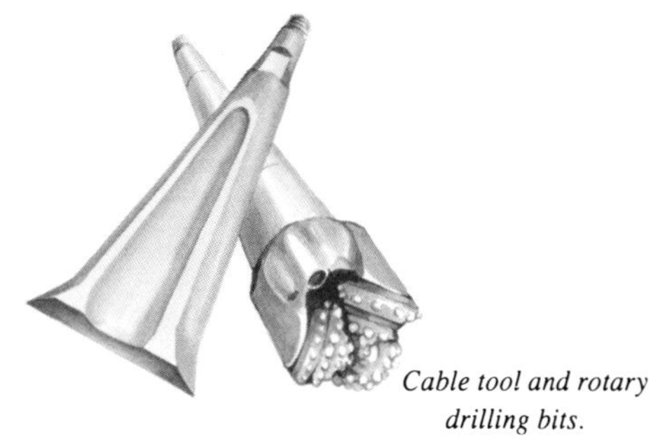

Cable tool and rotary drilling bits.

Edited by Susan F. Hodgson
Illustrated by Jim Spriggs

Foreword by George Deukmejian, Governor of California

Department of Conservation
Division of Oil and Gas

Library of Congress Catalog Card Number: 90-81141
ISBN 0-9627124-0-X

Published 1990
Printed in the United States of America

Published by the California Department of Conservation
Division of Oil and Gas, 1416 Ninth Street, Room 1310
Sacramento, California 95814

Division of Oil and Gas Publication TR40

All photographs attributed to William Rintoul are copyrighted, and all rights reserved. They may not be reproduced, stored in a retrieval system or transmitted in any form or by any means electronic, mechanical, photocopying, recording or otherwise, without prior written permission from Mr. Rintoul.

Photographs on pages listed below are used by the courtesy of and with permission from the following institutions:
Brea Historical Society 22, 28, 31
California State Library x, 4, 7, 9, 11, 16, 29
Kern County Museum 12, 13, 24, 33
Library of Congress 122, 124, 125
Long Beach Public Library and Information Center 5, 10, 13, 14, 15, 19, 20, 30, 40, 42, 43, 45, 46, 48, 49, 50, 60, 62, 64, 74, 89, 98, 99, 100
R. C. Baker Memorial Museum Inc., Coalinga 20, 21
Seaver Center for Western Research, Los Angeles County Museum of Natural History 3, 8, 9, 11, 48, 54, 55, 96
Security Pacific Historical Photograph Collection, Los Angeles Public Library 10, 47, 50, 53, 55, 56, 60, 79, 101, 135
The Huntington Library, San Marino, California 32, 33
West Kern Oil Museum, Taft 18

Contents

Author's Dedication — v
Foreword
George Deukmejian, Governor of California — vii
Preface
M. G. Mefferd, State Oil and Gas Supervisor — ix
1. Wood Derricks & Steel Men — 1
2. A Newcomer to the Oil Patch — 17
3. Shaping Up — 27
4. Decade of Discoveries — 41
5. The Great Depression — 57
6. The War Effort — 71
7. From Confusion Hill to the Subsidence Bowl — 81
8. Down to the Sea — 97
9. Full Steam Ahead — 111
10. Tapping Nature's Teakettle — 123
11. The Environmental Age — 135
12. At Work — 151
Division of Oil and Gas Managers — 165
Acknowledgments — 169
Further Reading — 171
Index — 173

This drawing of an oil field was published by the Division of Oil and Gas in its annual report from 1920 through 1953.

To the men and women who have
worked for the Division of
Oil and Gas during its
first 75 years.

OFFICE OF THE GOVERNOR
State of California

February 16, 1990

Dear Mr. Mefferd:

On behalf of the citizens of California, I am pleased to join in congratulating the Department of Conservation, Division of Oil and Gas, as it celebrates its 75th anniversary.

For 75 years, the division has effectively and capably served the citizens of California by supervising the drilling, operation, maintenance and abandonment of oil, gas and geothermal wells throughout our State.

The division's steadfast commitment to preserving and wisely using our State's natural resources has contributed to the growth and progress of our State while protecting our environment and the safety of our people. The dedicated men and women who have worked with the division throughout the years have certainly established a fine reputation for excellence and innovation and are to be commended for their fine record of service and achievement.

During this 75th anniversary year, I would like to thank the Department of Conservation, Division of Oil and Gas, for their many years of service to California and offer my best wishes for every continued success.

Most cordially,

George Deukmejian

George Deukmejian

Mr. M. C. Mefferd
State Oil and Gas Supervisor
Department of Conservation
Division of Oil and Gas
1416 9th Street, Room 1310
Sacramento, CA 95814

Division of Oil and Gas

Headquarters: 1416 9th Street, Rm. 1310
Sacramento, CA 95814

● Oil and Gas Offices

District No. 1	245 W. Broadway, Ste. 475 Long Beach, CA 90802
District No. 2	6401 Telephone Rd., Ste. 240 Ventura, CA 93003-4458
District No. 3	5075 S. Bradley Rd., Ste. 221 Santa Maria, CA 93455
District No. 4	4800 Stockdale Hwy., Ste. 417 Bakersfield, CA 93309
District No. 5	466 N. Fifth St. Coalinga, CA 93210
District No. 6	221 W. Court St., Ste. 1 Woodland, CA 95695

◉ Geothermal Offices

District G1	Headquarters Office
District G2	485 Broadway, Suite B El Centro, CA 92243
District G3	50 D Street, Rm. 300 Santa Rosa, CA 95404

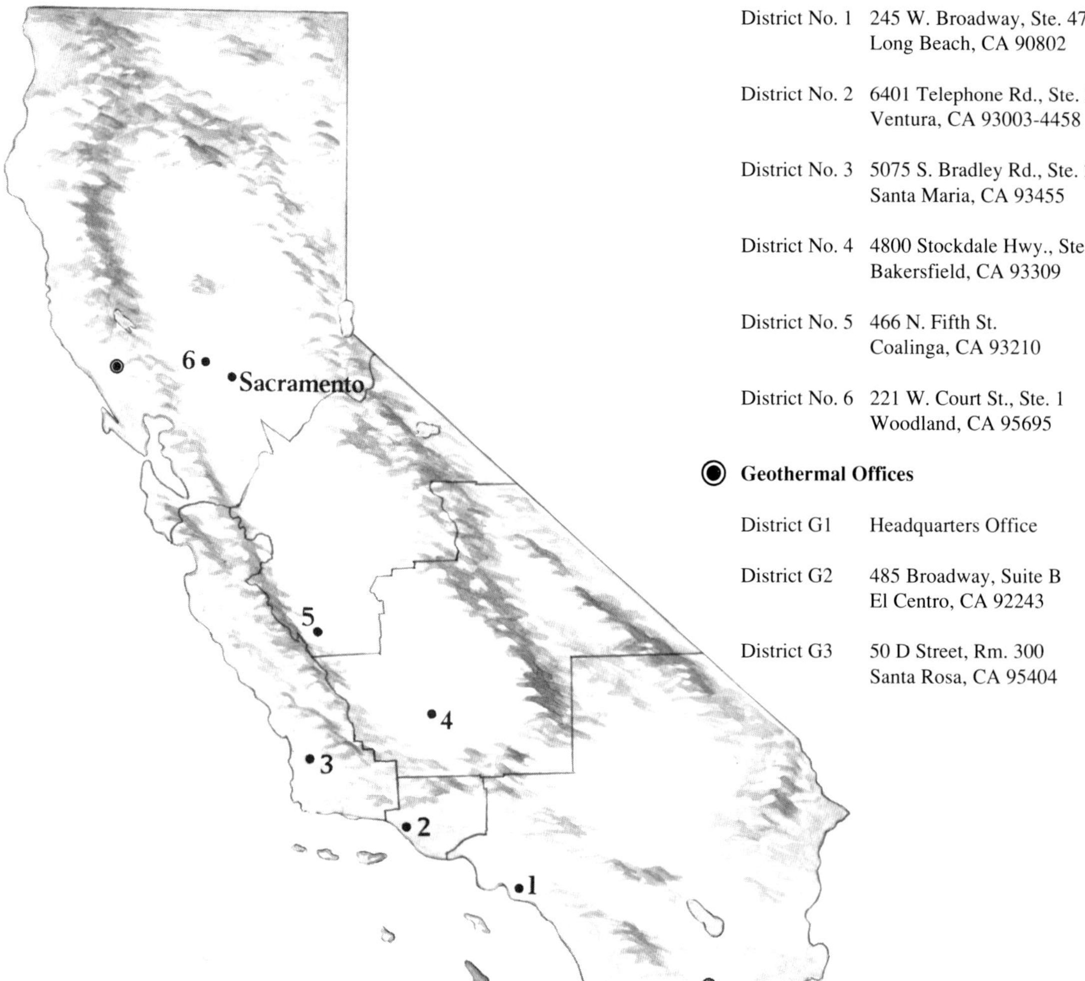

Preface

As a young geologist who came to the Division of Oil and Gas 31 years ago, I was trained by the people who, in turn, were trained by the very first Division employees.

Today, on the occasion of the Division's 75th anniversary, I am struck by this heritage, and by how unchanged the tone and substance of the Division's concerns and activities are.

Even though the Division and the industries it regulates operate in a much more complex environment than before, we still face the same questions-- where and how to drill and complete wells, and how to produce them properly. Although many of today's answers are different from those of the past, we see how modern choices have evolved from those made long ago.

These thoughts are confirmed by statements written in 1916 by Mr. McLaughlin, the first State Oil and Gas Supervisor, at the end of the Division's first year. He wrote, "Conditions contributing to successful work have been almost ideal, political interference has been entirely absent, (we have had) hearty cooperation by a large majority of the oil operators throughout the state, and last but by no means of least importance are the loyal and intelligent efforts of other officers of the department."

I believe the strength and success of the Division of Oil and Gas are due to the general good support we have received from different administrations, our willingness to work with the oil, gas and geothermal industries to resolve problems, and the many very professional and loyal people who have worked in the organization over the years. If we can maintain our flexibility, objectivity and dedication, we can continue to be a successful and effective regulatory agency.

Bill Rintoul captures this spirit in *Drilling Through Time*. The story he relates is a history of us all. I hope you enjoy it.

M. G. Mefferd
State Oil and Gas Supervisor
Sacramento, California
August 9, 1990

1 Wood Derricks & Steel Men

With high hopes and regional pride, Sacramento Oil Company joined the Kern River boom. (California State Library.)

On a frosty Wednesday morning four days before Christmas of 1933, more than one hundred unemployed oil workers climbed into the open ends of trucks to go to work along the shoreline where the Buena Vista Hills dip into the dry bed of Buena Vista Lake, seven and one-half miles northeast of Taft in the southern end of California's San Joaquin Valley.

Tule fog hung in a cold and dripping blanket over much of the valley floor as shivering men, who in better times would have been working in the West Side oil fields, began digging carefully into a sandy spit of land and nearby low-lying hills where a branch of the Yokuts American Indians named the Tulamni had lived centuries before.

The Great Depression was in full swing. The project was part of a government program conceived under the Federal Emergency Relief Administration to provide jobs through the winter in areas where there was heavy unemployment.

There was irony in the panorama that found skilled workmen who were unable to get jobs within the industry they knew best now engaged in a make-work project that would prove to be the most thorough investigation up to then of the California roots of that very same industry.

Through the next several months, men with shovels and screens uncovered the fragments of evidence that would enable the Smithsonian Institution not only to piece together a more complete picture of some of those humans who had been the first to make use of California's oil resource, but also to identify what might be regarded as a token of the prehistoric beginnings of the state's oil industry.

Through artifacts turned up by oil workers in the diggings they called the "shell bed" after its numerous freshwater shells and the "plum pudding" from the dark gray and formless soil, archaeologists learned that the Tulamni had used asphalt from seeps to bind together bundles of yucca fiber to make paint brushes, coated baskets with it and used it as a foundation for shell inlays and a cement in mounting chipped scrapers and cutting tools.

Among the findings were large balls of well-preserved asphalt. Archaeologists speculated that the "bundled" asphalt might have been stock used for trading purposes with tribes that inhabited regions lacking the abundance of seeps near the Tulamni mounds.

The Tulamni, of course, were not the only Native Americans to make use of the petroleum resource that seeped to the surface in California.

Their counterparts on the coast and as far north as the Mattole Valley in Northern California also harvested the sticky, dark material that made baskets watertight, secured arrowheads to wooden shafts and for some was said to have served as medicine for colds, coughs, burns and cuts.

Inevitably, the seeps from which American Indians obtained asphalt attracted the attention of those who pursued the frontier westward, inspiring efforts as early as the mid-19th century to make commercial use of California's petroleum resource.

Not surprisingly for a state that had achieved statehood through a gold rush in which immigrants poured in from all parts of the world, some of California's first enterprising oil developers tried to adapt mining methods to the production of the resource.

One such venture was the Asphalto operation in the foothills of the Temblor Range near what would later become the town called McKittrick.

A pair of entrepreneurs named Jim Hambleton and Judge Lovejoy organized the Buena Vista Petroleum Company in 1864 and proceeded to

At this natural seep, asphaltum flows down the slope and onto the beach, about one half mile east of Goleta Beach State Park, Santa Barbara County, once part of the Chumash Indian territory. The Chumash used asphaltum in many ways, as did many California Native Americans. (Division of Oil and Gas.)

This twined basket was waterproofed inside and out with asphaltum. It is from the collection of the State Indian Museum, Sacramento. (Jim Spriggs.)

Pahmit, a Dumna Yokuts, when he was about 105 years old. He watched San Joaquin Valley pioneers extracting asphaltum from the same deposits he and his family once mined. (From a 1929 photo by Frank Latta, in Handbook of Yokuts Indians.*)*

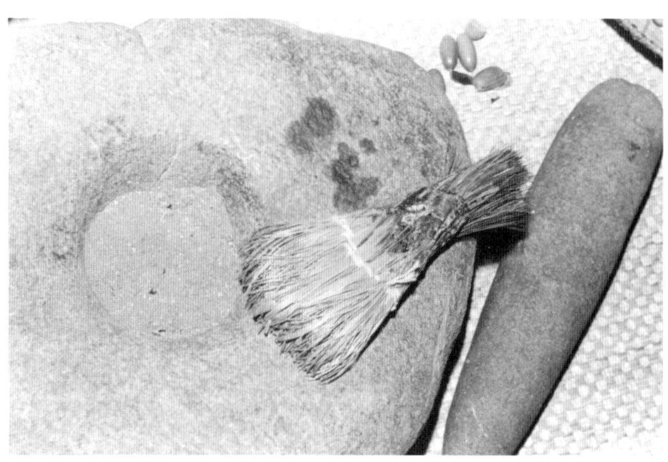

A Maidu acorn-meal brush, resting on a stone mortar filled with acorn meal, on exhibit at the State Indian Museum, Sacramento. The brush is made of soap-root fibers glued together with asphaltum and laced with string. (Jim Spriggs.)

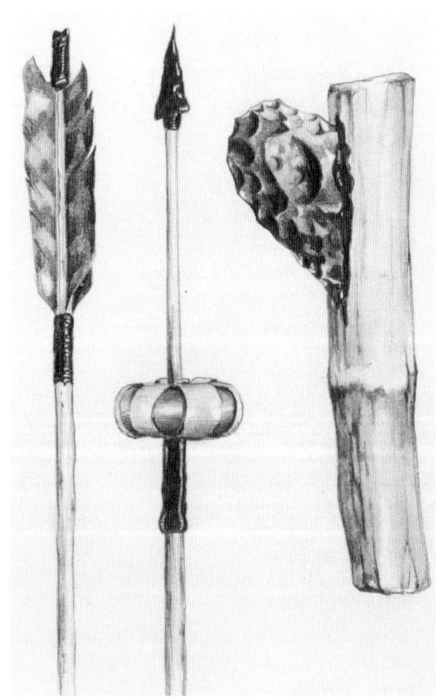

Yokuts knife and arrows made by Wahumchah, last surviving man from the Yowlumne tribe, using chunks of very hard, dry asphaltum. Softer asphaltum, when carried in a quiver or foreshaft pouch, would melt from the hot valley sun and the body heat of the hunter, gluing the arrows or foreshafts together. (From a photo by John Garcia in Handbook of Yokuts Indians.*)*

oversee the digging of shallow pits near an active seep. They built a still and set out to refine the tarry oil that collected in the pits.

Though their output was only a few thousand gallons of oil through the next year or two, the venture spurred others to try their luck.

At Asphalto, Mother Lode miners concentrated their efforts on ledges that yielded 80- to 90-percent bitumen, which was much better quality than the 34- to 35-percent asphalt produced on the island of Trinidad, then the world's main supplier of asphalt.

In open pits, the temperature in the summer rose as high as 140 degrees, limiting a work shift to no more than twenty minutes without relief. At some locations, asphalt was mined from tunnels and shafts more than 300 feet deep. From a depth of about eighty feet in one shaft, a column of pure asphalt ten feet high and six feet in diameter was shipped intact to San Francisco for public display.

Much of the asphalt was used for paving streets, some greased the skids for logs in timber country and some was used for sidewalks in San Francisco, where it brought a price as high as $30 a ton.

While Asphalto's developers largely ignored the oil that later would prove to be the real bonanza, a pair of San Francisco firms modified mining methods to produce crude oil near Ventura.

The Stanford Brothers and Hayward & Coleman chose as the setting for their experimental venture a steep, scrubby mountain named Sulphur Mountain, which rises abruptly north of the present town of Santa Paula.

Miners drove tunnels slanting slightly upward as deep as 400 feet into the sharply tilted strata of the mountain. The work was costly and also dangerous. Gas that collected in the tunnels was explosive.

Those who worked in the tunnels improvised a device known as a "water blast" to provide ventilation. The apparatus consisted of a wind sail attached to the top of a long, vertical pipe. The sail collected

Prospectors drilled near oil seeps like the La Brea tar pits, six miles west of Los Angeles when this picture was taken, circa 1906. Oil derricks in the distance share the skyline with trees in the then-bucolic setting. (Seaver Center.)

and directed air into the pipe, while water was simultaneously pumped in through an adjoining tube. The column of air and water cascaded down through the pipe into a box on the tunnel floor. The box separated air from water, "blasting" the fresh air through the shaft.

Completed tunnels generally were about five feet high and no more than four feet wide. Miners chiseled out a gutter on the tunnel floor and lined it with redwood planks. Because of the grade, oil and water washed easily down the gutters and out of the tunnels into holding tanks.

From five tunnels in Wheeler Canyon, Hayward & Coleman during their first year of operation ending in October 1866 recovered 2,570 barrels of crude ranging in gravity from 27 degrees upward to more than 30 degrees. The Stanford Brothers, in a slightly longer period, collected about 1,200 barrels from seven tunnels in nearby Saltmarsh Canyon. Crude oil from these and other tunnels was hauled by wagon to Ventura and shipped to San Francisco for distillation into lamp and lubricating oils.

While tunnels through the decade of the 1860s produced more oil in California than any other method, wildcatters early in the decade turned to the cable tool technology that had been borrowed by Colonel Drake from salt-well drillers in 1859 to drill a 69-1/2 foot deep well near Titusville, Pennsylvania. The Drake well had been hailed as the world's first oil well.

Two years after the Drake well, the first well drilled in California for the purpose of producing oil was sunk in Humboldt County. The well proved to be dry. Four years later, in 1865, Union Mattole Oil Company of San Francisco completed the first oil-producing well on the North Fork of the Mattole River, three miles east of the community of Petrolia.

Mother Lode miners in the 1860s drove redwood-shored shafts into California hillsides to help launch the state's oil industry. Above, remnants of an oil mine near McKittrick in 1956. (William Rintoul.)

At Asphalto, miners tapped ledges for high-quality bitumen that helped pave the streets of San Francisco and grease the skids for logs in timber country. (W. L. Watts. Donated to the California State Library by Frank Stockton.)

In June 1865, the company made its first shipment of oil, consisting of about 100 gallons, to San Francisco. There the Stanford Brothers refined and sold the light, greenish crude as "burning oil," reportedly for a price of $1.40 per gallon. At intervals over the next two years, Union Mattole Oil Company sent out larger amounts ranging up to 15 barrels.

A year after the first shipment of oil from Humboldt County, a wildcatter named Thomas R. Bard drilled several wells on the Rancho Ojai near Ventura, completing one designated as Ojai No. 6 for 15 to 20 barrels per day from a depth of 550 feet. The well was the first drilled for oil in California to yield extended production on a profitable basis.

Thanks to mines, tunnels and drilled wells, California had its first oil boom in the mid-1860s with some 65 oil companies operating from Humboldt County south to Ventura.

By 1867, the boom was over. With the end of the Civil War, crude production in Pennsylvania jumped more than 70 percent. The flow of crude through eastern refineries made a large volume of products available for shipment west, driving down the price of kerosene from $1.70 or more a gallon in 1865 to 54 cents or less two years later.

Well Pico No. 4, the first truly commercial oil well in California. (Long Beach Public Library.)

To make matters worse, the eastern products were superior. California kerosene burned with what was described as a "dull and smoky flame" and the lubricating oils contained no paraffin and flowed almost like water. Compounding the situation was the fact that California crude provided a much smaller proportion of refined products than Pennsylvania crude.

However, some activity continued. In Southern California's Pico Canyon, a San Francisco oil merchant named Frederick Taylor teamed up with a younger partner named Demetrius G. Scofield, who had gained a knowledge of oil operations in Pennsylvania, to purchase controlling interest in California Star Oil Works Company in the spring of 1875. The following year, the company completed Pico No. 4, producing 30 barrels a day from a depth of 300 feet. The well proved to be the first truly commercial oil well in California.

The performance led in the following year to the deepening of three wells that had been drilled earlier. Increased production demanded more refining capability. A small still that had been built in 1874 at Lyons Station, a stage stop nine miles from Pico Canyon, was dismantled and moved to a new location one-half mile east of Newhall near the Southern Pacific Railroad. Two larger stills shipped from Titusville, Pennsylvania, were added to the refinery,

Increased production from Pico Canyon wells in 1877 led to construction of the Newhall refinery by California Star Oil Works Company, a predecessor of Chevron Corp. (Long Beach Public Library.)

making it California's first commercial refinery.

Wildcatters fanned out in California through the remainder of the 1870s and the decade of the 1880s to drill and complete wells in Pico and Wiley Canyons near Newhall, at Moody Gulch in the Santa Cruz Mountains 50 miles south of San Francisco, at Adams Canyon near Santa Paula, in the Puente Hills of Los Angeles County and at Coalinga and in the Midway and Sunset districts of the San Joaquin Valley.

In the fall of 1892, a down-on-his-luck mining prospector named Edward L. Doheny drifted into Los Angeles from Colorado with less than $1,000 in his pockets and behind him

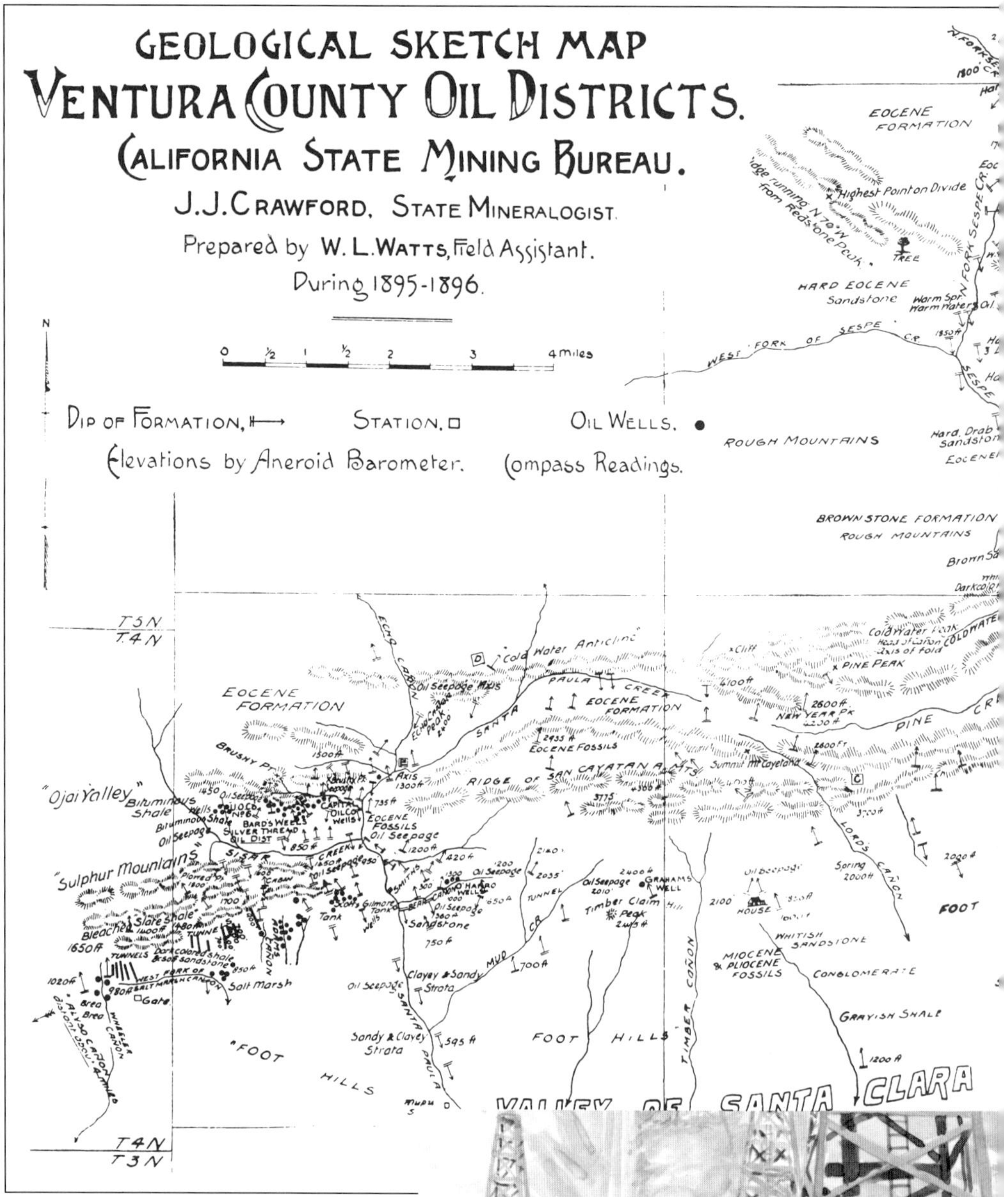

GEOLOGICAL SKETCH MAP
VENTURA COUNTY OIL DISTRICTS.
CALIFORNIA STATE MINING BUREAU.
J.J. CRAWFORD, STATE MINERALOGIST.
Prepared by W. L. WATTS, Field Assistant.
During 1895-1896.

0 ½ 1 ½ 2 3 4 miles

Dip of Formation. Station. Oil Wells.
Elevations by Aneroid Barometer. Compass Readings.

Rugged terrain often taxed the ingenuity of wildcatters. Above, remnants of a tramway that offered the only means of access to an early Ventura County well. The wooden derrick has collapsed against the hillside in back of the well. (Susan Hodgson.)

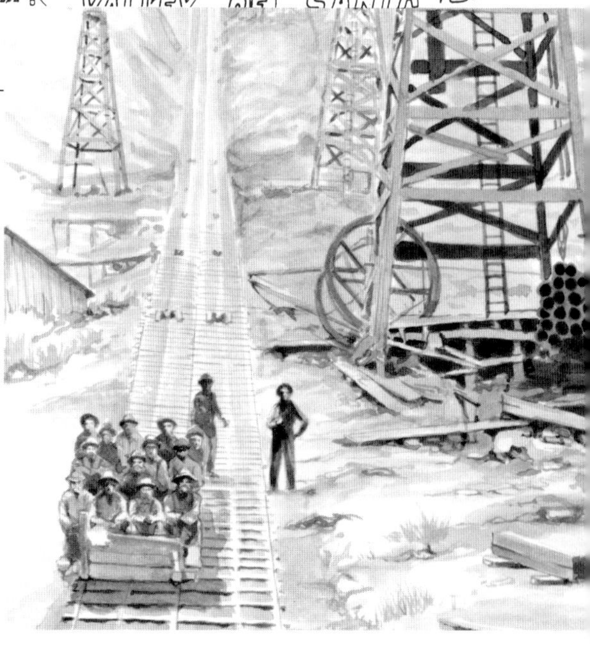

For some oil workers in Ventura County, going to work meant riding a tramway to drill sites inaccessible by roads.

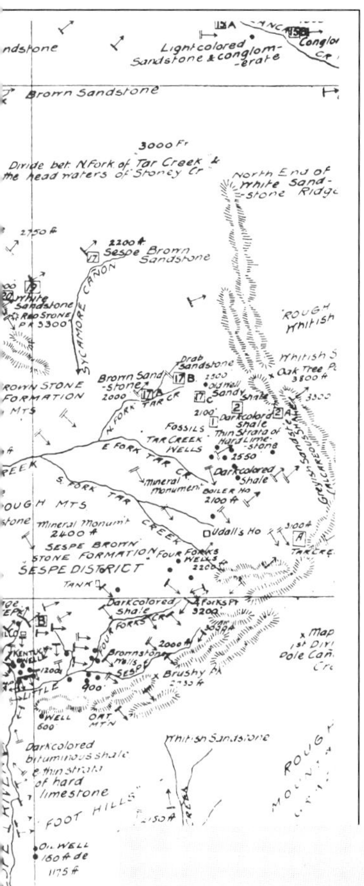

a singular lack of success in finding mineral wealth. About all the 36-year-old Doheny had to show for almost twenty years of combing the west for gold and silver were a scarred cheek and a mangled wrist that were the souvenirs of an attack by a mountain lion.

In Los Angeles, Doheny observed residents of the city gathering what they called "brea" from tar pits for use as fuel in coal-scarce California.

Realizing that the crude tar was petroleum that had congealed on contact with air, Doheny explored the residential neighborhood near Westlake Park, pooled resources with Charles A. Canfield, an old mining crony, and purchased a city lot for $400.

Unaware of oil drilling methods, they began by sinking a four-by-six-foot miner's shaft, digging it by hand with picks and shovels. They found an oil seep seven feet below the surface and kept digging despite the presence of gas. They finally gave up at 155 feet, nearly overcome by fumes.

Doheny then fashioned a crude drill from a 60-foot eucalyptus tree trunk and continued to bore the hole. On the fortieth day of work, gas burst out of the hole and oil bubbled up into the shaft.

Initially the well produced about seven barrels a day. The well was not the first to produce oil in Los Angeles, but the volume was an improvement, even if only a small one, over the two barrels a day that some other wells had tapped sporadically.

Soon after the first oil had been recovered, the Doheny well showed signs of improving with age. Before long, it was producing as much as 40 barrels a day, and the gravity was several degrees higher than the heavy crude that had been found in the past. And, there was another plus. The oil had been found closer to downtown Los Angeles, where industrial establishments offered a likely market. Even doubters began to see visions of a bonanza.

One more factor helped launch the Los Angeles oil rush as the greatest yet seen in California. The area in which oil was found had been subdivided into residential lots during a land promotion five years before. This meant that the entire area was up for grabs. If the owner of a

The Modelo Canyon area near Piru, shown in a photograph taken in 1898 by an unidentified California State Mining Bureau field assistant, was an early discovery. (California State Library.)

particular lot could not scrape together the money to drill a well, there were plenty of speculators eager to make a deal. Soon it seemed as if everyone who could promote the $1,000 to $1,500 it took to drill a well was competing for property. The boom was on.

The three years that followed the discovery of oil by the Doheny and Canfield well were ones of chaotic confusion and wonderful excitement in the booming Los Angeles City field.

With fortunes to be made, the residential district became crowded with promoters, drillers and derricks. Trampled gardens, chugging and wheezing pumps, flooded lawns and other nuisances went along with the attempt to turn backyards into pay dirt.

In its second year of production following completion of the Doheny and Canfield well, the Los Angeles City field produced 750,000 barrels of oil, pushing California's output past the one-million-barrel mark for the first time, and accounting for three out of every five barrels of the state's 1,209,000 barrels of production in 1895. By 1897, the area bounded by Figueroa, First, Union and Temple Streets contained more than 500 producing wells.

One of the most successful individuals to become a producer in the field was an unlikely candidate for the role of an oil operator. Mrs. Emma Summers was a piano teacher who was a native of Kentucky and a graduate of the New England Conservancy of Music. She had traveled to Los Angeles to teach music nine years before the oil discovery changed the course of the city.

One of her interests was real estate, and she was an occasional investor. Given the oil boom excitement that followed the Doheny and Canfield find, it seemed a logical jump from real estate into oil. Emma Summers enthusiastically took the leap, buying half interest in a well for $700. Before the well was completed, she had bought interest in a number of other wells and was in debt several thousand dollars. Unwilling to leave details to others, she personally oversaw the buying of tools and supplies, hired workmen and superintended drilling and producing operations. When the day's work was over, she hurried home to teach music, striving to earn more money so she might increase her investment in the oil field.

In addition to becoming an operator, she set out to corner the market on sales and succeeded to the extent of controlling more than half the production of the field before the turn of the century. At one point, she was selling more than 50,000 barrels of oil a month to downtown hotels, factories and The Pacific Light and Power Company, as well as to several Southern California commuter railroads and trolley lines. She soon became known as the "Oil Queen of California."

Drilling wells in the Los Angeles City field posed the problem of making oil production compatible with urban living. Noise, dirt, traffic, odors and waste disposal had to be dealt with. One solution proved unique. A homeowner had a rig in his backyard but no place for a sump in which to run the waste water and mud. However, his house had a basement, and that's where the mud went.

Los Angeles City field east from the corner of First Street and Belmont Avenue, circa 1900. (Seaver Center.)

The Los Angeles City field, spurred by a discovery by a down-on-his-luck mining prospector named Edward L. Doheny, gave Southern California its first great oil boom in the 1890s. (California State Library.)

By 1895, sturdy wooden derricks lined First Street in the Los Angeles City field. (Seaver Center.)

As wells proliferated, the price of oil dropped to as little as 10 cents a barrel. One day a producer with wells and a storage tank on the side of a hill by the intersection of Glendale and Beverly Boulevards was busy near the tank when a man came up and looked around. When asked what he wanted, the man said, "I'm from the City of Glendale, and we would like to buy a wooden storage tank, like this one."

The oil operator asked how much they would give for such a tank, and the man quoted a price.

The oilman's ears pricked up. He quickly offered, "What about this one right here?"

The man said, "Well, let me climb up and look at it." He climbed the ladder and examined the tank, then said, "This is about what we wanted and it would be just fine, except it's half full of oil."

"That's all right," the oilman said. "When do you want to come after the tank?"

The man said they probably could have the crew there the next day. "How soon would the tank be empty?"

"I'll empty it right now," the oilman said. Opening the valve, he let the oil run down the street.

While the Los Angeles City field was riding the crest as California's most productive field, the stage was being set for the San Joaquin Valley to assume the dominant role in the state's production picture.

The McKittrick field that followed the early mines and open pits of Asphalto opened a new chapter in 1896 when the Klondike Oil Company completed the Shamrock gusher flowing 1,300 barrels a day.

The Coalinga field, which had begun life in 1887 with completion of a 10 barrel-per-day well in the Oil City area, suddenly took on a new image with the completion of Home Oil Company's Blue Goose gusher in 1898 at a depth of 1,400 feet for a flow of more than 1,000 barrels a day.

In the spring of 1899, rumors circulated in Bakersfield that a couple of men poking around out on the Thomas Means place seven miles northeast of town had found oil.

Many were inclined to take the story with a grain of salt. Means had claimed for years that there was oil on his 20-acre property, but no one had taken him seriously. It was true there were some strange phenomena out that way. On John Barker's place near the Means parcel there was a spring from which natural gas escaped. Barker, in fact, had hung a bell-shaped trap over the spring, collecting gas that he piped to his house. However, the volume was barely enough for one jet. Also, from time-to-time a film of oil would be seen floating on the waters of Kern River, and there was talk that years ago a sheepman had used oil from a small seep to mark sheep. But a little oil floating on the river and a few sheep running around with tar-stained brands, a bubbling spring and a small gas jet burning in a remote farmhouse were a far cry from proving there was an oil field.

Some three weeks after the first rumor surfaced, a well-known and respected oil man named Angus Crites, who had worked for Jewett and Blodget in early West Side oil development near Maricopa, dropped into the offices of the *Daily Californian* in Bakersfield to advise the newspaper the rumor was true.

Crites said he had just come from the Kern River site, where he had seen whiskey barrels full of oil and oil in milk cans, kerosene cans and beer kegs, among other containers. He had watched men take about four barrels of oil from a hand-dug shaft, with the level remaining the same.

In the 1920s, the discovery of the Los Angeles City field was reenacted at the site of the E. L. Doheny well. (Los Angeles Public Library.)

On the floor of the discovery well, E. L. Doheny, left, recalled the Los Angeles oil rush and the new era it opened in the city's economic life. (Los Angeles Public Library.)

Loading oil into a tank wagon in Los Angeles City field for transport to market, circa 1900. (Seaver Center.)

The two men who had found Kern River's oil were Jonathan Elwood and his son, James Munroe Elwood, who ran a small wood yard in Bakersfield.

In a letter published in *California Oil World* in August 1911, twelve years after the discovery, Jonathan Elwood set down details of the find.

"James Munroe Elwood and I, Jonathan Elwood, alone and without assistance of anyone, discovered oil on the banks of the north bank of Kern River, seven miles northeast of Bakersfield on Thomas A. Means' farm," Elwood wrote to the weekly oil newspaper that had been founded in Bakersfield in 1908 by Charles P. Fox.

"This was in May, 1899. We made the discovery with a hand auger, under the edge of a cliff, close to the river. Our auger consisted of a piece of thin steel about four inches wide and twisted so as to bore a hole about three inches in diameter.

"We had a short piece of one-half inch iron rod, making the bit and rod together four feet long. A screw was cut on the end of this rod to receive a one-half inch gas pipe which we had cut in four and eight-foot lengths, so we could bore one and the other alternately and never have our auger handle more than four feet above the ground. We bored a number of holes 15 or 20 feet deep and every time would bore into water sand that we could not keep on our auger.

"We concluded that the bank must have slid down and that we were boring where the river had once been. We then went where the bank was worn off by the river perpendicularly 30 feet. We dug back into the bluff as if making a tunnel three or four feet, and set our auger on solid formation and in three hours we were in oil sand at a depth of only 13 feet. We had enough auger stem with us to go on to a depth of 25 feet and it was looking well.

"We then went up onto the bluff and commenced a shaft, and at the depth of 43 feet we again struck the oil sand. We were then obliged to get timber and curb as we went down, as the oil sand was too soft to stand up. We were obliged to put in an air blast to furnish fresh air to the man below on account of the strong odor of gas. At a depth of 75 feet there was so much oil and gas that we concluded we had better get a steam rig."

Horse-drawn tank wagons hauled Los Angeles crude oil to market. (California State Library.)

The hand-dug hole furnished the crude oil to power the steam rig, which drilled a separate hole north of the hand-dug hole. The drilled well went to about 260 feet and was completed on the pump for 15 barrels a day.

Almost overnight more than two hundred oil companies were formed to participate in the development of the Kern River field. Their names reflected everything from high hopes to pride in regional origins. Included among them were the Prosperity and the Blue Bird, the Pennsylvanian and the Hawkeye State, the Southern Cross and the Northern Light, the Sovereign and the Peerless, the Prince Edward and the Lord Roberts, the Apollo and the Aladdin, the American Eagle and the Uncle Sam.

During the first two years, the Kern River field saw the beginnings of development in an area comprising almost twelve square miles. As production climbed, the heavy crude found its major use as fuel for locomotives. The Southern Pacific Railroad promoted a sightseeing excursion from San Francisco, enabling tourists for a round trip fare of $10.60 to visit the Kern River field that powered the railroad's locomotives.

Four years after the discovery, the field's production had climbed to approximately 17 million barrels a year, enabling California in 1903 to become the nation's top oil-producing state, with an output of 24,382,000 barrels for the year. Kern River's share of

Hauling oil from the Kern River field, early 1900s. (Kern County Museum.)

the record output was seven out of every ten barrels.

Largely due to Kern River, California's production represented almost one-fourth of the United States' total production for the year. Closest competitors to the state were Ohio, 20,480,000 barrels, and Texas, 17,956,000 barrels. At the beginning of the century four years earlier, California's production of 4,325,000 barrels had put the state in fifth place behind Ohio, West Virginia, Pennsylvania and Indiana.

In the same year that California became the number one oil producer, one of the most important pioneering adventures in pipeline transportation was completed. It was a 280-mile, 8-inch line from the Kern River field to Point Richmond on San Francisco Bay, with a branch line from its trunk built into the Coalinga field to transport crude north to Standard Oil Company's refinery.

Construction of the pipeline was the first big job located any distance from Pennsylvania, which had been considered pipelining headquarters of the nation's oil industry because of its manufacturing facilities and supply of experienced labor.

The California pipeline was designed to transport thick, viscous oil, which was a type never before conveyed by pipeline over long distances. Such a feat was made possible by heating the crude oil and building pumping stations close together. From the start, the line was equipped with a complete telegraph system for rapid communication between the various pumping points and oil-dispatching headquarters.

Only one thing was missing from the Kern River field. The field could claim no gushers. The state had seen its first gusher in February 1892 when a Union Oil Company well in Adams Canyon near Ventura came in out-of-control, flowing an estimated 1,500 barrels a day. It was estimated the well produced 40,000 barrels before the flow ceased.

One-room schoolhouses sprang up in the oil fields, as at Sunset in 1902. (Long Beach Public Library, donated by Gordon P. Suiter, pictured in the first row at right.)

Almost overnight, more than two hundred oil companies were formed to participate in the development of the Kern River field, among them the San Joaquin and Kern Oil Companies, above. The woman in the foreground is unidentified. (Kern County Museum.)

There had been other gushers in the years that followed. One notable well was Union's Hartnell No. 1 at Santa Maria, which in December 1904 blew in with a roar that could be heard for a distance of four miles. The well flowed through 8-inch casing at a rate gauged roughly at about 12,000 barrels a day, and maintained the rate for about three months before a decline set in.

Other gushers that captured the attention of the California oil men were the Silver Tip No. 1 at Coalinga, which sprayed an estimated 10,000 to 20,000 barrels a day over the derrick top in September 1909, and Chanslor-Canfield Midway Oil Company's Midway No. 2-6 at Fellows in the Midway field that blew oil over the top of the derrick in November of that same year at a rate of 2,000 barrels a day. The latter was the first real gusher in either the Midway or Sunset districts, which had produced oil since before the turn of the century but had seen nothing like the Midway gusher before. About a year before the Midway gusher blew in, the stage was being set for another well 12 miles to the southeast, near Maricopa, that would take its place in the annals of California oil history.

The prime mover in the Lakeview venture was Charles Frederick Off, a native of Iowa who had begun his oil career in 1895 at the age of 29 drilling various wells on oil lands at Whittier and Orcutt.

Charles Off had turned his attention in 1908 to what would become the Midway-Sunset field. Opportunity presented itself in the form of Julius Fried, who had acquired some government mining claims near Maricopa but lacked the capital to do more than erect the wooden shacks required to hold the land.

In December 1908, Off was a key figure in the organization of Lakeview Oil Company, which was incorporated under the laws of California with capital stock of 2,500,000 shares at $1 each.

The newly-organized company purchased secondhand from Santa Maria Crude Oil Company an inventory of "machinery, tools and implements, pipe and pipe fittings, engine and boiler, tanks, a horse and buggy" for $11,548.52, paying with 18,130 shares of Lakeview stock. With high hopes, the company on January 1, 1909, spudded in to drill Lakeview No. 1.

Lakeview No. 1, California's greatest gusher, in the fourteenth day of what proved to be an 18-month flow of an estimated nine million barrels of oil. (Long Beach Public Library.)

Drilling dragged on, costing more than had been anticipated, and putting a strain on the company's finances. There was a showing of gas at 1,340 feet, but oil was absent.

A company with neighboring land and production provided some prospect of help. Union Oil Company, it developed, had its eye on the Lakeview land, not particularly for its oil potential but primarily as a site to build tanks to store crude oil for shipment through a pipeline the company was laying.

The Lakeview partners and Union entered into an agreement under which the latter, in return for the right to place tanks on the Lakeview land, 51 percent of Lakeview's capital stock and four-of-seven seats on the company's board of directors, agreed to continue drilling the Lakeview No. 1, then at 1,655 feet.

Union continued to drill the well along with three other wells the Lakeview crews had begun, but only on a spare-time basis when crews were available.

On the morning of March 15, 1910, drilling had reached a depth of 2,225 feet when a column of oil shot from the ground, setting in motion a progression of events that would destroy the derrick, flood surrounding land with oil and ultimately drive the price of oil down to 30 cents a barrel.

Nine days after the well blew in flowing at rates estimated as high as 125,000 barrels a day, *California Oil World* reported of the well, "It's hell, literally hell. It roars and rips like hell. It mounts, surges and sweeps like hell. It smells and terrifies like hell. It is as uncontrolled as hell. It is as black and hot as hell...Some of those who watched it the first night declared that it ejected glowing stones."

Three weeks after the well blew out, Charles Off addressed a postcard to his sister Julia M. Off in New York City. The picture on the card showed five men, one of them Off, standing by a pool of oil in which was reflected, from the background, the mighty Lakeview gusher, spouting oil high into the sky.

On the back of the card, at the top, Off wrote: "Mr. F. P. Dunlap, at extreme right, assisted me in organizing the company." Beneath, in the space reserved for the postcard message, he added: "The well is still flowing 42,000 bbls. daily. The greatest conquered oil well in the history of the world. We are saving all the oil. Have 250 men at work. Lovingly, Charles."

The Lakeview gusher flowed out of control for 18 months, finally stopping in September 1911 after it had produced an estimated nine million barrels of oil to earn enduring fame as California's greatest gusher.

"It roars and rips like hell," said one writer of the Lakeview gusher, described as flowing 48,000 barrels per day when this photo was taken. (Long Beach Public Library.)

2 A Newcomer to the Oil Patch

"Water, water, everywhere..." was the refrain for the Kern River field, which was one of the first major fields to suffer serious damage from encroaching water. (Elizabeth Hadden, California State Library.)

Even as California's free-wheeling wildcatters were firmly establishing the state as the nation's leading oil producer, trouble was brewing that threatened to make the oil industry's days of glory short-lived.

The culprit was water. If there were times that California seemed jubilantly afloat over a sea of oil, those who worked to recover the oil soon learned that the state also sheltered a bountiful supply of water, much of it saltier than the ocean.

When water broke into sand saturated with oil, the water displaced the oil, quickly turning what might have been a good oil well into a "wet" well, producing so much water the well was no longer commercial.

The problem posed by unwanted water seemed uniquely California's. In eastern fields, drillers found cap rock separating shallow water sands from deeper oil sands. They used the cap rock as a "ledge," setting casing in the hard rock, then drilling ahead in a smaller diameter hole into the oil sand. The casing effectively shut out water.

Unfortunately, California's oil sands too often lacked a strong shield of dense, impermeable rock. Water sands lay interbedded with oil sands. Frequently formations were poorly consolidated. Some were pulverized by the drilling operation, while others seemed soft as snowbanks. All too often water found its way into the oil sands, rendering what might have been a bonanza an economic liability.

The Kern River field, with its interbedded water and oil sands, was one of the first major fields to suffer serious damage. Wells were drilled close together. Operators differed on the proper depth at which to land the casing used to shut off water. Some strings of casing were landed in oil or tar sands, making it impossible to shut out water. Some wells went too deep, tapping bottom water that was not shut off properly. Because of the expense, some wet wells were abandoned without plugging between each sandy formation, allowing water to flood all the sands that had been penetrated.

The field scarcely had become the number one producing field in the state, by its rapid climb making California the nation's top oil-producing state, before an eastern visitor who had forged one of the most successful oil careers in the country sounded an alarm.

The visitor was the redoubtable Colonel John J. Carter, who had arrived in the United States at the age of five as a penniless orphan and at the age of twenty had won the nation's highest military honor, the Congressional Medal of Honor, at Antietam, and had come out of the Civil War as a brevet colonel. After the war, he had ventured into oil in Bradford, Pennsylvania, enjoying a high degree of success with The Carter Oil Company. He had sold a majority interest to John D. Rockefeller's Standard Oil Company, staying on as manager for the Carter operations.

In a report to Standard Oil Company in April 1905, Colonel Carter wrote after a reconnaissance of the Kern River field, "No matter what the production...was or what the hopes of the producer are, death and destruction surround that field, and it will only be a year or two at most, when it will be numbered with last year's snows and forgotten."

That very year, Kern River's production dropped from the previous year's record high of 17.5 million barrels to 14 million barrels, representing a decline of almost 20 percent. The invasion of water on the southern edge of the field forced the abandonment of many wells and the shutting-in of others, especially

on the San Joaquin, Peerless and Monte Cristo properties.

The Kern River field was not the only California field threatened by water. After visiting the Santa Maria field, Colonel Carter included it in his assessment of California fields that soon would belong to history.

In the battle to hold back the watery tide, operators turned to a variety of tools and techniques. One early approach involved lowering a bag of flaxseed into the hole to serve as a base upon which pipe, called casing, was landed. The theory was that the swelling of the seed on taking up water would hold back the water long enough to allow mud to settle and the formation to close in on the outside of the casing to create a permanent seal.

Some operators tried to seal off space between the "shoe" of the casing and the wall of the well with a mixture of clay and chopped rope. Some dropped rope and brick into the hole and tamped the mixture down to use as a foundation upon which to lower cement and lathe cuttings in tin cartridges. The cartridges then were broken up, allowing the cement to mix with water in the hole and form a plug. Others simply dumped in cement. In one Coalinga well, the operator, finding he had drilled into bottom water, plugged off the water with a conglomeration of "waste, rope, iron lathe cuttings and cement."

In the war against water, it soon became apparent that encroaching water did not recognize property lines. The operator who did not shut out water could damage not only his own wells but also the wells of neighbors. The situation was compounded by

One that got away, circa 1910 in the Midway field. (West Kern Oil Museum, Taft.)

Panorama of the now Midway-Sunset field with Standard Oil Company buildings, taken during the 1920s. Like Kern River, the field had water problems, often pitting one operator against another in a dispute over who was responsible for water damage to wells. (John O. Harries.)

every promoter who dropped out of sight without abandoning unsuccessful wells properly. Inevitably, there were confrontations.

On Twenty-Five Hill in the Midway field, one difference of opinion pitted Indian & Colonial Development Company against Traders Oil Company. The companies held adjoining leases, and the wells of both were plagued by water. The water was believed to be from bottom sand. Each company accused the other of not taking adequate measures to shut off the water.

After discussions failed to produce any satisfaction, Traders Oil Company proceeded to drill a new well to test each formation individually in an effort to identify the source of water. The company determined that the source was an intermediate water sand between the top oil sands and the bottom oil sands. It was this sand, not the bottom sands, that was the culprit.

Oil-drenched crew relaxes after successfully capping a Union Oil Company well in the Coalinga field, circa 1911. (Long Beach Public Library.)

Another dispute in the same field concerned a well on the Kalispel lease that produced large quantities of water. When the company proposed to case off the upper flooded oil-bearing horizon and deepen the well to lower sands that produced oil in wells on neighboring leases to the north and east, other neighbors to the south and west protested, expressing fear that the volume of water formerly pumped by the Kalispel well would back up and flood their top-sand producing wells. In the absence of well records for the area, operators could not agree on a solution.

In the Coalinga field, Ozark Oil Company faced off against Traders Oil Company over a Traders' well that Ozark claimed was producing one barrel of water for every barrel of oil, posing a threat to the Ozark property. Traders contended the well did not make any perceptible water. An investigation indicated the well was producing over 50 percent water, but Traders took no action.

As operators realized they could not solve the mutual problem of preventing water damage by simply exercising control over their own operations while being victimized by neighbors, the feeling grew that some sort of control would be necessary if California's oil industry was going to prosper and grow.

The first attempt to solve the situation by legislation came in March 1903 when the state, at the urging of concerned Kern River producers, passed a

A Pennsylvania machinist named Denny Driscoll who arrived in Coalinga in 1908 described one of the city's best-known attractions as "a solid block of saloons in a wide-open town that never closes down." (Long Beach Public Library.)

In the Coalinga field, as in other fields, entrepreneurs showed no lack of imagination in selecting names for their newly-formed companies. (Long Beach Public Library.)

law making it a misdemeanor to fail to shut off water above and below oil-bearing strata, or to abandon a well without securely filling the hole for at least 100 feet above the oil sands. The law provided no enforcement agency, which doomed it to failure from the start.

In March 1909, the earlier law was repealed and a new law passed requiring the owner to properly case and abandon wells. The new law provided for the keeping of a log and inspection by a county commissioner. The commissioner was to be appointed by county supervisors upon the request of three operators. The law also provided for filing and investigating complaints, and it declared any well not repaired a public nuisance. Enforcement proved lacking, sending a message that the state's most powerful producers were not yet ready to accept a law with teeth.

A third law relating to oil and gas was passed in March 1911. The act "to prevent wasting of natural gas" prohibited wasteful or unnecessary escape of gas into the air, and provided that such acts be treated as a misdemeanor punishable by a fine of not more than one thousand dollars or by imprisonment in the county jail for not more than one year, or by both such fine and imprisonment.

When the various laws failed to solve the problems, operators at Midway in the spring of 1912 moved to improve the situation by establishing the Kern County Oil Protective Association with an office in Taft. It was an advisory body, staffed by a deputy county commissioner and a geologist to whom the companies were to furnish well logs, records and drilling information.

The association received some support, but not enough to ensure success. Standard Oil Company, by now a major presence on the California scene, did not enroll, possibly for legal reasons, but the company

Oil companies provided comfortable homes at low rentals for married employees at lease camps. Here, a couple in a company house at California Oilfields Ltd.'s Coalinga Camp. (Frank Foster, R. C. Baker Memorial Museum Inc.)

did agree to supply the association with information at the company's command and to advance $150 a month to help defray the association's expenses.

One of the association's proudest achievements was the construction of the first complete peg model to give a three-dimensional picture of subsurface conditions as defined by existing wells. The model covered an area of approximately 20-1/2 square miles, extending over a distance some 7-1/2 miles long and 2-3/4 miles wide along the line of greatest development in the Midway-Sunset field from Fellows to Kerto, near Maricopa. Companies operating in the area bore the expense of building the model, which was exhibited at the Panama-Pacific International Exposition at San Francisco in 1915.

Two years after the Kern County Oil Protective Association was formed, a second association was set up at Coalinga. The Coalinga Water Arbitration Association was more than advisory. Its representatives were empowered to enter the property of any member to enforce a water shut-off. Among the members was Standard Oil Company.

Such a field-by-field progression might have continued, but operators who were able to agree to go as far as they had at Taft and Coalinga soon realized a better solution would be a statewide organization. They recognized, too, that some fields needing the benefits of cooperative action were too small to organize under the Kern or Coalinga patterns.

A regulatory agency operated by the state seemed to an increasing number of those associated with the oil industry as the best answer. One of those strongly supporting the concept was Fletcher M. Hamilton, the State Mineralogist, whose office had been created on April 16, 1880, by a law that set up the State Mining Bureau. From his vantage point as the top state official monitoring the oil industry, Hamilton concluded statewide action would be a better approach than piecemeal solutions.

Through the years, the State Mining Bureau had kept a relaxed eye on what was happening in the state's oil

The British corporation, California Oilfields Ltd., whose Coalinga property was purchased by Shell in 1913, sponsored soccer games for its employees at Coalinga. Wooden derricks provided a background of silent spectators. (Frank Foster, R. C. Baker Memorial Museum Inc.)

In the Olinda "field," later reclassified to become a portion of the Brea-Olinda field, new ideas for drilling methods were put into practice. These made cable-tool drilling successful in obtaining deeper production. Operators in the early 1890s experienced difficulty in getting casing below 850 feet until William Loftus in 1899 introduced the innovation of drilling with the hole full of water. By so doing, he was able to carry 5-5/8 inch casing to 1,465 feet, and the well flowed at a rate of 700 barrels per day. In the photograph, Olinda is seen circa 1910. (Brea Historical Society.)

fields. The Bureau's duties consisted principally of gathering and publishing information about the various mineral resources of the state rather than attempting to regulate the oil industry. In 1914, thinking that the time had come to take a closer look, Hamilton decided to make the first comprehensive survey of the California oil industry. He picked as the man to do the job a mining engineer named Roy McLaughlin, who had joined the Bureau in August of the preceding year.

McLaughlin's first introduction to oil had come unknowingly when, as a youth in a Colorado mining camp he had stood by the back of a traveling man's wagon and listened to the huckster extol the marvelous qualities of a potion he claimed would relieve many human ills and actually cure deafness. The huckster called the substance Romany oil. He sold it in pint bottles at a price of one dollar each, or three for two dollars. The Romany oil had a distinctive odor that McLaughlin would recognize later when he saw his first single-cylinder gasoline engine and smelled the odor of the fuel on which it ran, identifying gasoline and Romany oil as one and the same. McLaughlin had worked in mining camps at Bodie, California, and Manhattan, Nevada. When the mining boom burst, he had gone to work as a geologist for Associated Oil Company in oil fields in the Taft area. The survey he made of California's oil industry confirmed that in many fields damage to oil-bearing strata was caused by

Tool dressers cleaning boiler flues and fireboxes.

infiltrating water that had not been properly shut off.

Fletcher Hamilton, with the findings of McLaughlin's survey and McLaughlin's help, took the case for a statewide agency to oil producers in the field, laying the groundwork by first calling on the Coalinga Water Arbitration Association and the Kern County Oil Protective Association to seek their views and enlist their support. In its October 31, 1914 issue, *California Oil World* commended the approach:

"State Mineralogist Hamilton is going at the task in the right way by seeking up-to-date information and suggestions ... Hamilton deserves whatever assistance oil operators and landowners could find it within their power to give him."

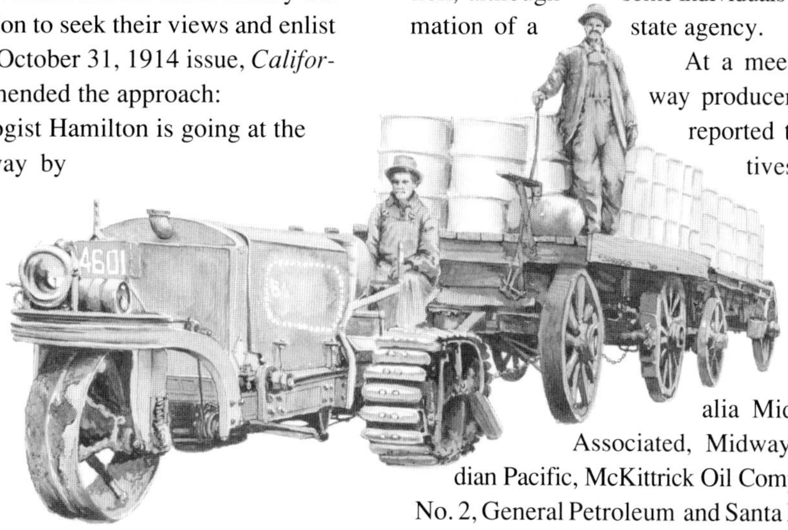

Some companies proved receptive to the idea, most notably Kern Trading and Oil Company. Another supporter was Standard Oil Company's Fred Hillman, who had come to Standard in 1911 from The Ohio Oil Company with instructions to build up production. Hillman had accomplished the assignment with great success, starting the company on the way toward becoming the nation's top oil producer before the end of the decade, largely from its successes in the San Joaquin Valley's Midway field and in the Los Angeles Basin on lands acquired from Murphy Oil Company in the Coyote Hills and at East Whittier. Hillman, a practical oil man who had come up through the ranks, concluded state regulation was the only sound way to police fly-by-night operators.

In a swing through San Joaquin Valley fields to explain the proposed "water law" to producers, McLaughlin reported he found no organized opposition, although some individuals opposed the formation of a state agency.

At a meeting with Midway producers, McLaughlin reported that representatives from Indian & Colonial, C. C. Harris Oil Company, W. O. Maxwell, A. J. Pollok's Miocene, Visalia Midway, Fairfield Associated, Midway Supply, Canadian Pacific, McKittrick Oil Company, Lakeview No. 2, General Petroleum and Santa Fe endorsed the concept. He noted, however, that those offering support were field men, not top management.

By December, McLaughlin had enlisted help in preparing a bill from a committee with representatives from the Independent Oil Producers Agency and, one each, from the Kern County Oil Protective Association, Coalinga Water Arbitration Association and Orange, Santa Barbara, Los Angeles and Ventura Counties.

The full text of the proposed bill "to protect oil lands menaced by water" was released on January 2,

Supply wagons entering the outpost oil camp of Reward in the McKittrick field, circa 1913. (Ann Bass McDonald.)

1915. The measure was introduced in the State Legislature on February 6, 1915, sponsored by the State Mining Bureau. It was estimated that the cost of providing protection against water damage through the new agency to be established by the bill would be less than one mill per barrel of oil produced by each operator. The measure passed unanimously.

The law establishing the Department of Petroleum and Gas of the State Mining Bureau became effective August 9, 1915. It provided that the Department should be under the general jurisdiction of the State Mineralogist, who should appoint as Supervisor an engineer or geologist experienced in the development and production of petroleum. It would be the duty of the Supervisor to supervise the drilling, operation, maintenance and abandonment of petroleum or gas wells in such manner as to prevent damage to the petroleum and gas deposits of the state from infiltrating water and other causes.

Wells could not be drilled without first notifying the Department, including in the "Notice of Intention to Drill" the precise location of the well as measured from a corner of the section or property, the derrick floor elevation, the estimated depth at which the water shut-off was to be made and the estimated depth at which productive oil or gas sands would be found. Notices of intention to make a water shut-off test, deepen or redrill an existing well or abandon a well also had to be filed and the proposed plans reviewed by the Department before work could begin.

While some functions of the Department were advisory, in cases where it found that water had not been shut off satisfactorily, it could order the operator to do such remedial work as the Department believed necessary, subject to review at the operator's request by a board of fellow operators.

State Mineralogist Fletcher Hamilton appointed Roy McLaughlin to be the first State Oil and Gas Supervisor. In a spirit of cooperation, the Kern County Oil Protective Association and the Coalinga Water Arbitration Association turned over their records to the state agency.

In a general statement directed at operators, McLaughlin described the information operators were required to furnish the Department of Petroleum and Gas as falling into two categories. "The first is a complete record or log of each well, giving in detail each and every step taken in its construction and repair, as well as the location and thickness of all strata penetrated so far as can be determined.

"The second class of information consists of a record continually brought down to date and showing the amounts of oil and of water produced by each well."

On the keeping of oil well logs, McLaughlin commented, "In most lines of business an accurate inventory would show exactly what had been obtained in return for the outlay. Many oil operators, however, spend their development funds and have little or nothing in the way of records to show them whether or not the work has been done properly or economically."

To remedy the situation, McLaughlin suggested operators see to it that the operations of the drilling crew be completely written down each and every tour, or shift.

"One of the handiest methods of keeping these daily drilling reports is to have them bound in book form, one page for each tour," McLaughlin advised. "A carbon copy of each page should be made, promptly removed from the book and filed in a safe place, because the original book is subject to loss or easily

The first published reports in 1911 of an oil discovery "10 miles northwest of the Nacirema well on the McKittrick Front" were vague, not only about the extent of the find but also the operator's name, initially calling the successful wildcatter "Bell Ridge," then in a later edition, "Belle Ridge." The lettering on the side of a truck delivering casing made it plain the company was Belridge Oil Company. (Kern County Museum.)

becomes soiled and illegible."

The daily drilling report was to show the depth of the well at the beginning and the end of the tour, what sort of work the crew was engaged in and the size, weight and amount of casing put in or taken out. It was also to record the depths at which there were changes in formation, describe or name the formations and state what evidence there was to indicate the presence of oil, gas or water in the well.

"The work of the State Mining Bureau in protecting the oil fields against infiltrating water depends upon full and complete logs of wells and the law requires that they be furnished to the bureau," McLaughlin reminded operators.

To assist in the regulation of the oil industry, McLaughlin appointed as his first deputies M. J. Kirwan (in Coalinga), Chester Naramore (in Taft), Robert B. Moran (in Los Angeles) and Roy E. Collom (in Santa Maria).

The Department of Petroleum and Gas was to operate on a yearly budget of $45,000. The industry the Department was charged with overseeing was producing approximately 237,000 barrels a day of oil from some 7,000 wells. McLaughlin estimated the total investment in the industry to be at least $250 million, including "...80,702 acres of proved oil land with a market value of at least $1,000 per acre, or a total of $80,702,000; approximately 2,000 miles of pipe lines costing on an average of $20,000 per mile, or a total of $40,000,000; nearly 30 refineries, with a total daily refining capacity of about 175,000 barrels of crude oil and representing a total investment of probably $15,000,000; approximately 40 tank steamers serving the California oil fields, having a total carrying capacity of about 1,500,000 barrels and costing not less than $15,000,000; and 7,000 producing oil wells that cost over $100,000,000."

"The members of the staff keenly appreciate the responsibility resting upon them," McLaughlin wrote.

In an editorial opinion of the role State Oil and Gas Supervisor McLaughlin and his deputies were to play, *California Oil World* in its December 11, 1915, edition sounded an optimistic note:

"The unanimous passage of the law placing the regulation of drilling of oil and gas wells under the State Mining Bureau, and more especially its subsequent widespread approval among oil operators, serves as proof that Californians are fully alive to the importance of protecting their oil deposits from waste.

"If the law served no other purpose than to merely relieve the operator of unjust public suspicion, it would be worth while. The law follows an entirely new plan and is attracting widespread attention in other oil-producing states. From the manner in which the administrative work of the State Mining Bureau is already being aided by oil men it is no exaggeration to say that within the next year, when the aims of the law become fully understood, no other state in the Union will be taking as much care of its oil resources as is California."

The well marked by the wooden derrick to the left of the first office of the Department of Petroleum and Gas in Taft occupied a unique place in California oil annals. In late 1910, Standard Oil Company had landed 16-inch stove pipe casing at 545 feet and just begun to drill ahead when the fishtail bit became cocked over in the hole. When the crew was unable to recover the bit, a wiry crewman named John Stuck volunteered to go down the hole to kick the bit loose. First, a flame was lowered into the hole to make certain there would be air for Stuck to breathe. Fortunately, they had not cut any gas pockets. Stuck removed his jacket and allowed the crew to tie a line under his arms. They lowered him down the hole, feet first. On bottom, Stuck, barely able to breathe, managed to draw up his legs enough to kick the bit until he had straightened it, then tugged the line to let the crew know he was ready to come up. Gingerly his fellow workmen pulled him out, not caring to lose another "fish," particularly a human one. On the surface, Stuck said he had figured it would take "some hard kicks—and so it did." (Division of Oil and Gas.)

3 Shaping Up

In Bakersfield on January 9,1917, Department of Petroleum and Gas personnel posed for a family portrait after one year and five months of operation. Standing, left to right, G. McGregor, inspector, Bakersfield; Murray, driller; R. N. Ferguson, deputy, Taft; I. M. Johnson, office deputy, San Francisco; J. H. Dougherty, inspector, Taft; R. E. Collom, deputy, Santa Maria; and R. D. Bush, inspector, Taft; seated, left to right, M. J. Kirwan, ranking deputy, Coalinga; R. P. McLaughlin, supervisor, San Francisco; and R. B. Moran, deputy, Los Angeles. (Division of Oil and Gas.)

Through its first year, the Department of Petroleum and Gas used what Roy McLaughlin in his first annual report as State Oil and Gas Supervisor described as a "uniformly lenient" policy of administration.

McLaughlin made it clear the agency would prefer to accomplish desired results through persuasion and, perhaps, disarming flattery.

"In making this first report," McLaughlin wrote, "it is pleasant to state that conditions contributing to successful work have been almost ideal."

There were words of praise for almost all concerned. "In the first place," McLaughlin declared, "the widest possible latitude has been afforded me by the State Mineralogist, Mr. Fletcher Hamilton, and it is of especial public interest to be able to state that political interference has been entirely absent."

There were more laudatory words for operators. "The second factor is the hearty cooperation by a large majority of the oil operators throughout the state," McLaughlin wrote.

And for the staff. "Last but by no means of least importance," McLaughlin stated, "are the loyal and intelligent efforts of other officers of the department: Messrs. M. J. Kirwan, Chester Naramore, Robert B. Moran and Roy E. Collom, who have served as deputies in the various oil fields, and Mr. I. M. Johnson, who has attended to the clerical work of the San Francisco office, which was of great proportions, due to the commencing of an entirely new department."

In defense of the lenient approach, McLaughlin said there might be some who would criticize such a policy, "...but it must be remembered that ultimate success of the protective work depends upon a thorough and sympathetic understanding of its details by the owners of wells and their employees, who must be depended upon to ultimately carry out most of our suggestions.

"Such cooperation could not have been obtained by merely applying the penalties provided by law, without first explaining the constructive methods provided by it."

The upbeat tone carried at least a hint that the picture might not be completely rosy. "The leniency of the past year will not, however, be indefinitely continued," McLaughlin advised.

Though commending operators for cooperation, McLaughlin had less sanguine observations to make about the workings of the oil industry.

"There is probably no large business so inefficiently conducted as is that involved in the production of oil in California," McLaughlin wrote.

"The annual losses, due to unsystematic work and actually paid out of pocket, amount to hundreds of thousands of dollars. Bankruptcy would speedily follow such management in any line of business, not dependent upon either a most abundant natural supply of crude material or fresh infusions of capital from other sources. These two alleviating conditions

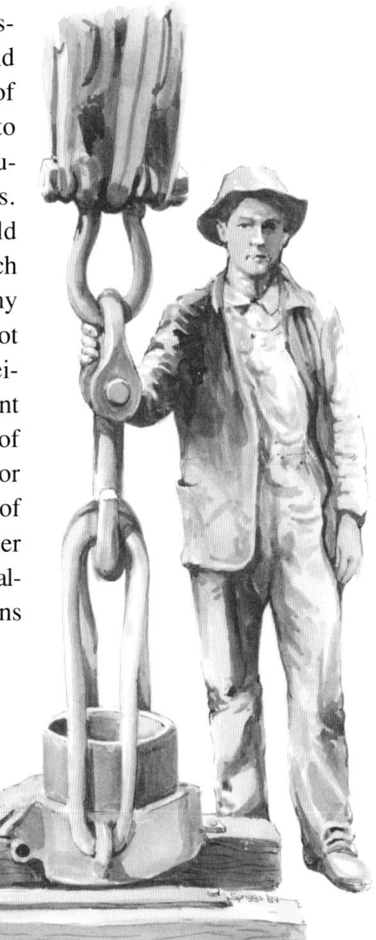

can not be expected to continue indefinitely," McLaughlin concluded.

The annual budget of $45,000 for the Department of Petroleum and Gas, McLaughlin made plain, was inadequate. In the Santa Maria field, it had been noted that a large amount of water came from a limited number of wells, indicating the possibility of remedying conditions.

"It was our intention to make a number of experiments with dyes introduced into wells under such conditions that results might be carefully observed at neighboring wells," McLaughlin wrote. "If properly carried on, such tests must certainly yield valuable results. The present extremely high cost of all dyes makes such work impossible on any extensive scale."

The lack of funds had a severe impact on Deputy Supervisor Chester Naramore's district. The district, covering McKittrick, Midway-Sunset and Kern River, was the most active of the four districts. The number of new-well notices for the first year was 273, or almost twice as many as the 141 for the Coalinga, Lost Hills and Belridge district and nearly four times more than the combined total of 71 for the Los Angeles area: the district including Orange and Ventura Counties had 59 notices, and the district including Santa Maria and outlying areas had 12.

For Naramore working out of Taft, it quickly became apparent that it would be necessary to have an assistant stationed in Bakersfield to handle the routine work of the Kern River field. McLaughlin concurred with Naramore's call for help in December 1915, four months after the Department of Petroleum and Gas had commenced operations, but was unable to secure the state funds necessary to hire an assistant. The Taft office found itself unable to handle Kern River work except for formal water shut-off tests.

Even in this regard, the district office had to

CLASSIFICATION AND WAGE SCALE OF OIL FIELD EMPLOYEES

Union Oil Company of California

Occupation	Daily Wage Rate Effective July 1, 1919
CABLE TOOLS	
Driller	$8.75
Tool Dresser	6.25
Circulator or Third Man	5.25
ROTARY TOOLS	
Driller	8.75
Bit Dresser (Only when bits are dressed in rig)	6.75
Derrick Man	6.25
Rotary Helper (Dressing bits in rig)	5.75
Rotary Helper (Not dressing bits in rig)	5.25
RIG BUILDERS	
Head Rig Builder	8.75
Rig Builders	7.75
Rig Builders Helpers	6.75
WELL CLEANERS	
Head Well Cleaner	6.75
Well Cleaner Helper	5.25
WELL PULLERS	
Head Well Puller	6.25
Well Pullers	5.25
ENGINEERS, FIREMEN AND PUMPERS	
Engineers	5.50
Firemen	5.00
Pumpers	5.00
Oilers	5.00
DEHYDRATOR OPERATORS	
Dehydrator Operators (Employed continuously in operation of large retorts)	5.50
Dehydrator Operators (Operating in plants in conjunction with other work; classified as pumpers)	5.00
ROUSTABOUT CREW	
Head Roustabout	6.00
Roustabouts (Not entitled to 25c contingent pay)	5.00
GASOLINE EXTRACTION PLANTS	
First Class Plants	
First Engineer	5.75
Second Engineer	5.50
Large Booster Stations	
Engineer	5.50
Small Booster Stations	
Engineer (Operating booster station with or without other work)	5.25
Firemen, Oilers, Pumpmen, Traptenders	5.00
FIELD EMPLOYEES	
TEAMSTERS AND TRUCK DRIVERS	
Teamsters (2 horse) (Not entitled to 25c contingent pay)	5.00
Teamsters (4 horse)	5.00
Teamsters (6 horse and over)	5.25
Light Truck Drivers (Up to and including 1 ton) (Not entitled to 25c contingent pay)	5.00
Light Truck Drivers (Over 1 ton and up to 3 tons)	5.00
Heavy Truck Drivers (3 tons and over)	5.75
Stablemen (Split time understood; eight hours work within any 10½ consecutive hours) (Not entitled to 25c contingent pay)	5.00
BOILER WASHERS	
Boiler Washers (No chipping; employed continuously) (Not entitled to 25c contingent pay)	5.00
Boiler Washers (Where chipping is required, employed continuously)	5.25
GENERAL	
Warehouse Yard Men (Checkers)	5.00
Tank-car Loaders (Not entitled to 25c contingent pay)	5.00
Roustabouts (Not entitled to 25c contingent pay)	5.00
GARAGE REPAIR MEN	
Head Garage Repair Man	6.50
Garage Repair Men	6.00
Garage Repair Men's Helpers	5.25
STEAM, GAS ENGINE AND PUMP REPAIR MEN	
Repair Man No. 1	6.25
Repair Man No. 2	5.75
Repair Man's Helper	5.25
FIELD ELECTRICIANS	
Electricians (Journeymen) (Not entitled to 25c contingent pay)	7.00
Electricians Helpers	5.25
FIELD SHOPS	
Acetylene Welder	6.50
Heavy Fire Blacksmith (Employed principally in tool manufacture and at heavy forge work)	8.00
Light Fire Blacksmith	6.50
Other Blacksmiths	6.00
Blacksmith Helper (Heavy Fire)	5.50
Blacksmith Helper (Light Fire)	5.25
Heavy Hammer Driver (2500 lbs. or over)	5.50
Light Hammer Driver (Under 2500 lbs.)	5.25
Boilermaker	6.50
Boilermaker's Helper	5.25
Machinist No. 1	6.75
Machinist No. 2	6.25
Pipe Machinist (Six inches and over)	6.00
Pipe Machinist (Under six inches; at bolt machines and other work)	5.75
Shop Helpers	5.25

Also all employees in the classification of roustabouts or upward in the above schedule, employed between the dates of December 31st, 1918, and June 30th, 1919, will receive fifty cents per day back pay for such time worked,—excepting where employees have received an increase in wages between these above dates, such increase will be deducted from the amount of back pay. Employees can make written request for back pay either by letter or on forms furnished by the company and such request must be made during the year of 1919.

Within forty-five days after June 30th, 1920, the company will pay to each employee classified in the above schedule (excepting those marked "not entitled to 25c contingent pay") an additional twenty-five cents per day for each day worked between June 30th, 1919, and June 30th, 1920. Such payment will not, however, be made to employees who have not worked at least thirty days for the company or to any employees who are discharged for wilful misconduct or who may leave the service through concerted action such as strikes, walk-outs, etc.

The board allowance made by the company will be discontinued as of June 30th, 1919, and where the company operates the Boarding Houses, the rate will be $1.25 per day.

Drillers and head rig builders topped the payroll of oilfield employees with a daily wage of $8.75. The lower end of the pay scale was reserved for such classifications as firemen, pumpers, oilers, teamsters, roustabouts, boiler washers and stablemen. (Brea Historical Society.)

The advent of World War I in Europe spurred the demand for oil, and with it an accelerated search in California. Despite the confidence of the man atop the derrick, the wildcat, near Galt in Sacramento County, proved to be a dry hole. (California State Library.)

seek outside help. A volunteer named C. E. Ballagh graciously accepted an appointment as special deputy supervisor and gratuitously gave the Department of his time, when he could spare it from his regular duties, to make shut-off tests on numerous wells in the Kern River field when the deputy was unable to make the trip from Taft. Ballagh was superintendent of the Amaurot Oil Company, Apollo Oil Company and Four Oil Company in the Kern River field.

Another field where the Department wanted to undertake work was the McKittrick field, which Naramore described as having a very serious water problem.

"As soon as the personnel of the Taft office is increased to the point where clerical work does not monopolize most of the time of the deputy," Naramore reported to McLaughlin, "cross sections will be constructed, water conditions will be tabulated, and then with the assistance of the operators, an effort will be made to locate the offending wells."

On April 20, 1917, twenty months after the Department of Petroleum and Gas had come into being, the agency's work took on added importance with the entry of the United States into the World War that had engulfed Europe since August 1914. In the months that followed, the nation's oil-producing industry with California in the forefront as the number one producer had the task of supplying the oil that was vital to the conduct of the war.

Within weeks, the state's oil fields bristled with armed guards and patriotic fervor swept the fields. In Taft, the *Daily Midway Driller*, the town's newspaper, reported that practically all the companies on the West Side had put on double guards to protect oil storage tanks. On Section 1, the newspaper said, Standard Oil had put in what the paper described as the "greatest searchlight in the field, a light whose rays reach three miles." The light was backed up with a machine gun mounted on a turret, the paper reported, noting that guards would treat all trespassers as spies. William Jennings Bryan came to town on a tour urging people to get behind President Wilson in the conduct of the war.

The camaraderie among the men who manned the rotaries that began replacing cable tool rigs in the 1910s and early 1920s often was shared by their families. Above, a group described by one of those present as "part of our happy family" at a rig working in the Devils Den area south of Coalinga. (Esther Baisden.)

At a hearing in Taft hardly more than a month after the United States entered the war, numerous oil company representatives appearing before a two-man governor's commission composed of Max Thelen, railroad commissioner, and David M. Folsom, Stanford professor, complained of the

To the field men, the oil company camp in the Lost Hills field was known as "the Oasis." The camp hummed with activity under the impact of the war in Europe. (Long Beach Public Library.)

difficulty of getting casing, particularly 10-inch pipe. A spokesman for Chanslor-Canfield Midway Oil said his company had ordered 200,000 feet of casing but not a foot of the pipe was on the way.

Against a background of shortages of men and material in the oil fields, Roy McLaughlin noted in his second annual report as State Oil and Gas Supervisor, "The entrance of the United States into the war makes the saving of oil extremely important at the present time."

Perceiving a need for what he described as "the utmost economy of labor and materials, together with the maximum production," McLaughlin concluded it was time for a public appraisal "showing to what extent the large operators are utilizing the latest and most approved methods."

In the report, McLaughlin said operators producing more than one million barrels of oil per year might be classified in three groups: first, those using thoroughly organized and competent technical departments to study underground conditions and direct development work; second, those using technical assistants to some extent, but in an inadequate and imperfectly coordinated manner; and third, those having no organized department to take advantage of technical information.

Those singled out for praise included Southern Pacific Company, formerly operated as Kern Trading and Oil Company, which McLaughlin described as "the pioneer in systematic or scientific oil field work in California, having used such methods from the time of its first operations nearly ten years ago;" and Shell Oil Company, "...which has been operating in California for some three years, and has from the first utilized scientific methods to good advantage."

Others mentioned by McLaughlin included Standard Oil Company, which "readily adopted suggestions made by the bureau, and a technical department has been installed, with every evidence of

becoming most effective;" General Petroleum Corporation, which "has for some three or four years directed its underground work by means of technical aid;" Honolulu Consolidated Oil Company, which "had been specially zealous in developing and utilizing technical and scientific methods;" Atchison, Topeka and Santa Fe Railway Company, operating in the oil fields as Petroleum Development Company and Chanslor-Canfield Midway Oil Company, where "engineering investigations have to some extent governed development work in one of the fields (the company produced oil in three fields) and recently arrangements have been made to so conduct all work;" and Union Oil Company, which "has recently informed this department that technical methods are to be adopted throughout the territory."

Of the E. L. Doheny interests: American Petroleum Company, American Oilfields Company Ltd., Pan American Petroleum Company and Doheny-Pacific Petroleum Company, McLaughlin wrote, "So far as we are informed, after diligent inquiry, technical methods have not been systematically utilized."

Most operators producing less than one million barrels of oil annually had not applied modern methods, McLaughlin said. However, he added, there had been notable efforts made by some small concerns to direct their development work scientifically, naming Alaska Pioneer Oil Company, Interstate Oil Company, Lakeview No. 2 Oil Company, Miocene Oil Company, Montebello Oil Company, Nevada Petroleum Company, Pacific States Petroleum Company, Recovery Oil Company, Riverside Portland Cement Company, Santa Maria Oil Fields, Ltd., and Universal Oil Company.

There were signs by the end of the second year that the "lenient" approach mentioned prominently in the first annual report was wearing thin.

In the Coyote Hills field, Providential Oil Company failed to file log records and monthly production reports. A number of requests were made for the records and various blank forms providing spaces for required information were sent to officials of the company. After waiting patiently for more than a year, the Department made a formal demand for records. The company failed to reply or file any records, and the deputy supervisor swore out a criminal complaint. After a summons had been served, the company immediately filed the records, and the complaint was dropped.

In the Montebello field, the Department resorted to legal action against Red Star Petroleum Company in regard to the company's Baldwin No. 4 well. The complaint charged that the operator had not taken samples as recommended to determine the depth at which to cement the string of casing to protect the first oil zone from water, and subsequently had cemented the casing 80 feet below the depth where the oil-bearing formation was encountered.

The Brea Canyon "field," later to become part of the Brea-Olinda field, was an 1899 discovery that, like other California fields, was called on to make its contribution with increased production during the first World War. (Brea Historical Society.)

At the trial, the three drillers testified they had received instructions while working on another well under the same management not to report any oil or gas showings on the log. One driller testified that acting under direct instructions from the foreman he had used a five-gallon can to remove evidence of crude oil from the rotary ditch prior to the arrival of the representative of the Department. The Department said it would not have given approval for the cementing of casing at the approved depth if it had known the procedure would shut off 80 feet of oil-bearing formation. After hearing evidence for some four days, the court dismissed the case.

The second of two legal cases initiated by the Department involved a complaint filed against Thomas A. Slocum, due to the failure of Slocum & Company to file logs of wells drilled on its property in the Santa Paula field and for failure to file monthly production reports, notices of intention to drill new wells, notices of intention to abandon, redrill or deepen old wells and for failure to notify the Department for water shut-off tests before completing new wells.

After repeated efforts, in writing, on the part of the state deputies to obtain the necessary information and the disregard by the company of these requests, the complaint was filed by the District Attorney of Ventura County. The defendant entered a demurrer, which was not sustained. The company in the meantime prepared all of the desired records and submitted them to the Department. The complaint was dropped.

A continuing theme sounded by the Department of Petroleum and Gas through the beginning years was criticism of companies, both large and small, for failure to recognize the important roles to be played by engineers and geologists in the development and continued health of the California oil industry.

Acceptance by field men of the new breed of college-trained specialists often was lukewarm, if not hostile. Geologist Ralph Arnold, left, and H. R. Johnson in field outfits, near McKittrick, 1908. Arnold carried a camera, binoculars and a field bag for rock samples. Johnson added a canteen. (The Huntington Library.)

"While discussing the importance of systematic work and its value to oil operators themselves, entirely aside from legal requirements of this department," McLaughlin wrote in the *Second Annual Report*, "it may be well to explain the functions of an engineer or geologist, which seem to be understood by comparatively few oil men."

McLaughlin conceded that most companies were beginning to recognize the usefulness of geologists in choosing new or prospective oil lands, but charged that too few companies realized the importance of continuing to use the services of a geologist or engineer after development had begun. "The larger operators...will particularly benefit by considering this question," McLaughlin wrote.

A year later, an obviously frustrated McLaughlin sought to enlist the support of public opinion, which he described as eventually controlling the methods of all important lines of business and industry.

In the *Third Annual Report*, the State Oil and Gas Supervisor wrote, "Three large oil concerns have not yet adopted modern scientific methods of directing their drilling work, namely: Standard Oil Company, Union Oil Company of California, and a portion of the so-called Doheny interests."

Standard drew fire on the basis of a study of 50 of the company's wells in the Whittier field. M. J. Kirwan, deputy supervisor, said the study showed comparatively few wells in which certain zones were

College-trained geologists and engineers began to play an increasingly active role soon after the turn of the century. Of the picture of himself at work in Santa Barbara County in 1906, geologist Ralph Arnold wrote, "Near view of flinty Monterey (Middle Miocene) shales on Sisquoc River 6-1/2 mi. northeast of Sisquoc..." (The Huntington Library.)

A General Petroleum Corporation field headquarters office, San Joaquin Valley, circa 1919. (Kern County Museum.)

A field geologist's home away from home. Of this picture, Ralph Arnold wrote, "View of Arnold and Anderson's room in Arthur Hotel, Lompoc, showing boxes of fossils and general chaotic condition of the apartment. October, 1906." (Ralph Arnold, The Huntington Library.)

known to be protected, that is, where tests had demonstrated that water was not entering from either above or below the zone. The majority of wells, the report said, fell in the "doubtful" class, where a shut-off had been made below the zone without demonstrating that it was protected.

Many of the log records were described as "very incomplete." In the case of one well, the Department had made a recommendation that a zone thought to be water-bearing between the first and second oil sands be tested. No reply was received. "The manner in which the company ignored this letter suggests that more positive action than merely writing letters seems necessary in matters of this kind," Naramore wrote.

Union was criticized by Roy Ferguson, deputy supervisor in Taft, for cementing a string of 10-inch casing about 270 feet above the probable position of the top oil sand in the International No. 7 well on Maricopa Flat. "The present shortage of oil, as well as casing, gives cause for serious reflection before the issuance of an order which either postpones the production of oil or requires an increase in the amount of casing to be used," Ferguson wrote. "On the other hand, the protection of the oil-bearing deposits is of prime necessity, particularly at the present stage of national necessity." Three months after completion, Ferguson said the well was producing considerable water and "has actually depreciated the value of the entire property."

Doheny was cited for deliberately withholding information from the Department on drilling activity in the Montebello field and for failing to follow a recommended program of frequent sampling which resulted, in the case of at least one well, of cementing oil sand behind pipe.

Industry's attitude often touched on the continuing battle between those who had learned in the field and those who brought knowledge gained in universities, pitting college-trained engineers and geologists against field personnel who had little formal education but had long experience in the field.

McLaughlin had encountered the industry's suspicion of college-trained geologists some years before when, after first working as a geologist in the gold mining industry, he went to work for Associated Oil Company in the San Joaquin Valley. Initially he traveled about in a light wagon with an assistant who served as cook and camp helper—and was paid more money than McLaughlin.

No living quarters were provided for McLaughlin and his wife at the main company lease, and the couple was forced to rent rooms in such small houses as were available. Most of the drilling crews and their families lived on the lease in what was called "Rag Town," a collection of tent houses.

McLaughlin was given no definite instructions other than to collect well logs and histories, draw cross sections and construct underground contour maps. When he was transferred to Taft, he was provided with a Locomobile that frequently developed engine failures at remote places, requiring the driver to walk several miles for mechanical repairs. The duties of his job were vague and such recommendations as he made, suggesting the acquisition of land, seemed to produce no results. Concluding that the job had no future, McLaughlin had resigned in favor of setting himself up as a consultant in San Francisco. The move turned out to be a wise one, for the company soon after abolished its geological department.

In outlining the general requirements governing

members of the Department of Petroleum and Gas, McLaughlin made it plain where he stood in the matter of college-trained employees vs. those who had come up from the field. "It may be necessary for some men, who have had considerable so-called 'practical' experience in the oil fields, to actually unlearn or forget some theories which may have appeared to them as facts. In reality there are too few known facts relative to the conditions with which we have to deal. A statement that a proposition should be considered in a certain way, merely because it has been so considered in the past, has little or no force in this department," McLaughlin concluded.

Even as McLaughlin was propounding the Department's philosophy, the stage was being set for a confrontation with the largest producer in California. The setting was a low range of hills in the southern end of the San Joaquin Valley where sagebrush grew and weeds turned brown in summer. The hills were named the Elk Hills after the tule elk that once had roamed the area.

Standard Oil Company in 1909 had purchased a section of land at Elk Hills known as the "school section" because it had been granted to the State of California to aid in development of the school system. There had been no production at the time Standard acquired the land. The company held the section for speculation.

In the meantime, the same speculation that had brought the oil company to Elk Hills also attracted another party, though not for precisely the same purpose. The U. S. Navy, recognizing the need for a supply of crude oil to fuel oil-burning ships, had asked Congress to set aside possible oil-bearing land in the public domain to serve as a reserve. Congress had done so. In September 1912, President Taft had issued an executive order creating Naval Petroleum Reserve No. 1 from some 38,000 acres at Elk Hills, including the school section, though not challenging Standard's ownership of the land.

For six years, it remained to be seen whether the U.S. Geological Survey, acting for the Navy, or Standard Oil, for that matter, had made a good choice. Late in 1918, Standard moved in to drill the first wildcat on the school section. In January 1919, the well flowed 256 barrels a day from 2,532 feet, proving up the Elk Hills field. (Three successful wells drilled in Elk Hills in 1911 by Associated Oil Company were not considered to have done so.)

Standard moved rapidly to develop the field. The company's operations fell within the district of the Department's Roy Ferguson, a geologist.

After a detailed study of the ongoing Standard operation at Elk Hills, Ferguson became convinced of the existence of a gas zone in the locality being developed and recommended that casing be cemented at a considerable depth above the area where Standard was developing the oil sand. The company paid no heed to the recommendation and managed to drill the second and third wells through the gas without recognizing its existence, getting a 400 barrel-per-day well with the second well and an 850 barrel-per-day well with the third.

An increasingly nervous Ferguson watched the development. When Standard proposed another well that summer, he enlisted McLaughlin's support. Ordinarily McLaughlin left final decisions to the field deputy, but in this case he joined Ferguson for a meeting with Standard officials and the two state men insisted that the next well should land casing to protect the gas zone. However, the state law lacked teeth to enforce the recommendation.

After an angry discussion, McLaughlin and Ferguson agreed to make no further protest to Standard's plans for the next well so long as the well after that would be drilled in accordance with their recommendations. The compromise was accepted by Standard's manager of production, who assured McLaughlin and Ferguson in no uncertain terms that Standard Oil Company knew how to drill wells and that after completion of the agreed-on wells, the company would expect no further suggestions from them.

Early one morning in June, thereafter, McLaughlin was on the night train bound for Bakersfield where he was to inspect a well that was approaching critical depth. He raised the shade in his Pullman berth and saw on the distant horizon a plume of flames and smoke rising from Elk Hills. He knew without being told that nature had seconded the caution he and Roy Ferguson had urged on Standard. The Standard well, Hay No. 4, had gotten away from the drilling crew at around 2,000 feet and was flowing an estimated 30 million cubic feet per day of gas. It took some four days for the company to bring up the dozen boilers necessary to generate the columns of steam required to smother the fire. Afterward, the well was capped and, when lines were ready, put on production supplying gas for the City of Los Angeles.

Nature was not yet through emphasizing its point. Around midnight on Saturday, July 26, on the school section, Standard was bailing at a depth of 2,140 feet at the Hay No. 7 when the well blew out with a thunderous roar. Gas quickly caught fire, burning fiercely in a torch that could be seen from fifty miles away.

It was an awesome sight. Men who only a few weeks before had battled the flames at the other Hay well began preparations to contain the new gasser. Flames shot three hundred feet into the

Flames from Hay No. 7 turned night into day at Elk Hills in the summer of 1919. (E. W. Brubaker.)

sky. It appeared they were being fed by substantially more gas than the earlier well. Standard rounded up all available men and began moving in boilers to mount an attack. It was estimated gas was escaping at a rate of fifty million cubic feet per day.

The towering flames attracted attention throughout the fields and in the neighboring towns of Taft and Bakersfield. On the following day, which was Sunday, motorists flocked to see the spectacle, clogging bumpy oilfield roads and making more difficult the task of moving in and rigging up boilers. Standard assigned some 125 men to crews laboring to control the wild well.

The *Bakersfield Morning Echo* said of the spectacle, "It is the largest flame ever seen on the West Side. The roar of flames can be heard in Taft, nine miles away. The western sky at night from Bakersfield is alight with the mass of fire that spreads over the horizon. The volcano shoots about three hundred feet in the air. Flames are about three feet wide at the base but explode in fan shape as they shoot into the air so the great pillar of fire is probably twenty-five feet wide at its greatest diameter. The heat is intense and one can not come closer than three hundred yards."

The burning gas created an inferno in which workmen approached the flames like knights stalking a dragon, moving toward the flames behind corrugated iron shields twelve feet in height and twenty-five feet wide. They wore woolens to protect themselves. The team effort included men farther back who manned hoses pouring a constant stream of water on the men who labored to set up boilers and rig lines so that steam might be turned against the well.

By Tuesday, the third day, ten boilers ringed the gasser. The signal was given. Ten jets of steam were played on the fire. However, the steam failed to make any impression on the huge column of burning gas.

By Friday, the sixth day, sixteen boilers ringed the site. The *Bakersfield Morning Echo* reported,

State-of-the-art well control technology featured five-foot-long torpedoes, each filled with 100 pounds of dynamite, used to snuff out fires. Left to right, Ford Alexander; Clarence Hendershott, Alexander's brother-in-law; and K. T. Kinley. Kinley's sixteen-year-old son, Myron, not shown, sometimes worked with Alexander and the others and later became the oil fields' premier tamer of wild wells. (Boyd Alexander.)

"The large tongue of flame has caused curious coloring around the mouth of the well. The ground has turned peculiar shades of red and gray. Some state that it is chemical action caused by the awful continuous blast of heat and others that the heat alone has caused change on the earth."

That night the steam from sixteen boilers was hurled at the monster. Again, the steam proved ineffective. The fire appeared to be increasing in volume, and the call went out for more boilers and chemicals. A baggage car full of carbon tetrachloride was dispatched hastily by rail from San Francisco.

On the Sunday that marked the beginning of the second week of the fire, wool-clad workmen hit the stubborn flames with steam from twenty boilers. Twelve pumps mixed fifteen tons of carbon tetrachloride into the steam. Flames sank low into the crater, then, with a roar shot high into the air.

By this time, the flames that lit the night sky had begun to lose appeal as a novel oilfield spectacle and were beginning to assume proportions of a menace. There was some thought the earth might cave in, leaving a vast and yawning cavity where gas had been contained. Fear was expressed the well would deplete the supply of gas for the towns of California, plunging them back to the days when coal was used to heat homes.

Standard turned to Ford Alexander, an oilfield dynamiter who only recently had returned to Taft from the boom at Ranger, Texas. Alexander directed workmen in tunneling as near to the mouth of the fire as possible. On Monday, the ninth day, charges of dynamite were placed in the tunnel. The dynamite was set off. The explosion failed to close off the mouth of the well, though it did alter the hole. Flames continued to leap into the sky, though now they were only half as high. Alexander prepared another attack. He decided dynamite, to be effective, would have to be discharged not from a tunnel but from directly over the hole. To get the dynamite in place, he strung a cable from a wooden derrick some distance from the wild well and anchored the cable across the draw to a post implanted in the ground. A trolley was fitted to the cable so that the trolley could be drawn into position over the hole.

After the blazing well was brought under control, Standard Oil Company's Hay No. 7 was hailed as the world's most productive gas well. (John C. Maher, United States Geological Survey.)

Dynamite was attached to the trolley. The first attempt to work it into position was made that night, the same Monday the tunnel charges had proved ineffective. Wheels of the trolley tangled in the cable. The car was hastily withdrawn with its load of dynamite still intact. Alexander waited for daylight to make the next try. A decision was made to hit the gasser simultaneously with the dynamite blast and with all the steam and carbon tetrachloride twenty boilers could muster.

Shortly after ten o'clock on Tuesday, the tenth day, more than one hundred workmen manned positions at boilers, hoses, pumps and cable. Dynamite was brought skillfully into position and detonated, even as steam and chemicals poured into the crater. The dynamite blast cut off the pyre of gas. Flames burst in the air. Then, the well fell silent. Cheers broke from the workmen. A moment later rocks and dirt began blasting from the crater, but the accompanying gas did not catch fire. Men immediately began clearing the ground to begin the effort to cap the well.

Sixteen days later on Friday, August 21, workmen had reached the casing and dug a large hole around it to position equipment to end the uncontrolled flow of gas. Twenty-six days after it had begun, the flow of gas from Hay No. 7 ended. Standard officials estimated the rate of flow at the time the well was capped at 140 million cubic feet per day. Gate valves were installed and pipelines laid to put the gas well on production.

During the next seven years, the well produced an unprecedented forty-three billion cubic feet of gas. It was hailed as the world's most productive gas well. For the Department of Petroleum and Gas, it represented a vindication of judgment.

4 Decade of Discoveries

In the wake of Shell Oil Company's discovery, the mansion that Andrew Pala had built as a retirement home on the crest of Signal Hill became bachelor quarters for the stream of employees the company brought in to develop the field. (Long Beach Public Library.)

As the United States enthusiastically bounded into the decade of the "Roaring Twenties," a shortage of oil loomed on the horizon.

If no other event had existed, the recently concluded World War had demonstrated convincingly the importance of oil. American companies had supplied 80 percent of the oil used by the Allied forces, enabling them, in the words of Great Britain's Lord Curzon, "to float to victory on a wave of oil." As plain as it was that petroleum products would become even more important, it seemed to many observers just as clear that oil was not as abundant as might be desired.

In July 1920, State Oil and Gas Supervisor Roy McLaughlin took note of the situation in an article in the Department of Petroleum and Gas' monthly publication that since April 1919 had superseded the annual report. Now, monthly increments were incorporated at the end of the fiscal year into one volume that sufficed as the annual report.

"Recent demands in California for petroleum products, particularly fuel oil and gasoline, forcibly demonstrate that the supply is limited," McLaughlin wrote. "The development of various industries on the Pacific Coast will in many cases force reliance upon other sources of power than petroleum."

McLaughlin cited statistics showing the state's maximum production had occurred in 1914 with slightly more than 300,000 barrels of oil per day, followed by a decline in the following year leading to a fairly constant production for a period of about four years in the neighborhood of 260,000 barrels per day.

The most important feature of the statistics, McLaughlin continued, was the average daily production per well. During the preceding five years, the average output of an oil well in California had fallen from about 40 barrels per day to 30 barrels daily.

"In order to obtain any considerable increase in the total oil output of the state," McLaughlin concluded, "it will be necessary to drill an extraordinarily large number of new wells in proved territory or to discover new fields giving extraordinarily large production in new wells."

Fortunately, the California oil industry seemed confidently positioned to meet the challenge. Through the preceding decade, industry leaders had learned, sometimes painfully, that a company could overlook the contributions of geologists and engineers only at risk of being left behind by competitors.

The acceptance of college-trained men and the challenge they—and the industry—faced had had its impact. In the fall of 1919, Dr. Bailey Willis, head of the geology department at Stanford University, had quoted oil production, consumption and reserves data to his class in structural geology and concluded

by telling them, "The world is at your feet."

The comment by Willis did not fall on deaf ears. Enrollment in the 1920 field geology class totaled sixty-one students, an all-time high for the university.

As the bumper crop of geologists and engineers joined those who had sought California's new fields in years past, an increasing number turned their attention to a longer look at the Los Angeles Basin.

Before the dawn of the 1920s the basin had seen its share of exploration and discoveries, including the proving up of seven giant fields, that is, fields that would produce 100 million barrels or more of oil. The fields and their dates of discovery were Brea-Olinda, 1880; Beverly Hills, 1900; West Coyote and East Coyote, 1909; Montebello, 1917; and Richfield and Santa Fe Springs, 1919. Only the most optimistic geologist could have predicted what was going to follow in the '20s.

Standard Oil Company started the decade of discoveries with the completion of a wildcat at Huntington Beach in June 1920. The initial production at Huntington A No. 1 was an unspectacular 70 barrels per day of 20-gravity oil. In the next several months, the company paid bonuses of some $200,000 for leases on about 900 acres belonging mostly to Huntington Beach Land Company. The amount of money led to speculation that the Standard discovery might be better than the initial well indicated. In November of that same year, the company brought in a follow-up well making more than 2,000 barrels daily. The fast-flowing well touched off a stampede for leases.

Huntington Beach had been laid out into townlots by developers envisioning a land boom. Most of the lots had been sold before the oil discovery. Because of the widely dispersed ownership and the small-sized individual holdings, the field quickly became the setting for a townlot drilling campaign that saw operators rushing to "capture" the oil beneath not only their small parcels but those of neighbors as well.

Inexperienced operators, stock-promotion schemes and haphazard drilling methods made the work of supervision difficult. In November 1921, R. E. Collom, who had succeeded Roy McLaughlin as State Oil and Gas Supervisor, charged that the development of the Huntington Beach field "by no means measures up to the standard set by the Department of Petroleum and Gas for the orderly development of an oil field."

"In some instances at Huntington Beach," Collom wrote, "cooperation with this Department was willfully lacking. In most instances cooperation was lacking simply because the operator did not know what steps should be taken to drill a well properly. A great deal of educative work was necessary, as there were more inexperienced operators drilling at Huntington Beach than in all the rest of the oil fields in California."

J. B. Case, the Division's Chief Deputy, added, "Many wells were drilled to a considerable depth, some of them even having the water string landed,

At Huntington Beach, derricks and bathers marched to the sea in the 1920s. (Long Beach Public Library.)

before a notice to drill was filed with the Department. When the operator was asked to file the necessary notice and attention was called to the fact that notices had not been filed as required by the law, the excuse was generally offered that that matter had been left to the contractor. It indicates a complete disregard for the state law governing oilfield operations."

In examining tour reports used in the field, Case said many were found to be inadequate. "Generally only carbon copies were available at the rig," Case charged. "In a number of instances, these copies could not be read because poor carbon paper had been used."

He noted, too, that elevations furnished with notices to drill were not always reliable. "In many cases, they are not even good guesses," Case wrote. He cited as an example the elevations furnished for two wells only 100 feet apart. Huntington Sure Shot Oil Company gave an elevation of 114 feet for its No. 1 well; Huntington National Oil Company for its No. 1 well gave an elevation of 35 feet. When a scientific measurement was made, it developed the elevation for the first well as 49.32 feet, for the second, 47.56 feet.

However, two developments helped the Department weather the storm. The workload was now divided into five districts from the four that had been established in 1915. The original four districts had operated from offices in Coalinga, covering the Coalinga, Lost Hills and Belridge fields; Taft, covering the McKittrick, Midway, Sunset and Kern River fields; Los Angeles, covering activity in Los Angeles, Orange and Ventura Counties; and Santa Maria, covering the Santa Maria field and outlying districts.

With the revision, the workload was divided into District 1, covering the Counties of Los Angeles, Riverside, Orange, San Diego, Imperial and San Bernardino; District 2, covering Ventura activity; District 3, covering the Counties of Santa Barbara, San Luis Obispo, Monterey, Santa Cruz, San Benito, Santa Clara, Contra Costa, San Mateo, Alameda and San Francisco; District 4, covering the Counties of Tulare, Inyo and Kern; and District 5, covering the Counties of Fresno, Madera, Kings, Mono, Mariposa, Merced and all other counties in California not included in any of the other districts.

Additional assistance came in the form of pay raises, authorized by the legislature, that boosted salaries for all but the State Oil and Gas Supervisor, whose pay remained at $500 a month. The Chief Deputy's pay was raised from $333.33 per month to $416.67. Other pay raises were for the five deputies, $300 to $375; four petroleum engineers, $200 to $250; twenty one inspectors, $170 to $200; twelve office assistants, $100 to $150; one chief clerk, $150 to $200; and four draftsmen, $125 to $175.

While Standard Oil Company and others concentrated their efforts on Huntington Beach, Shell Oil Company was taking another look at nearby

Big Four Oil and Gas Company of Fallon, Nevada, was incorporated under the laws of the State of Nevada to participate in the oil boom. (Long Beach Public Library.)

Signal Hill. A company geologist named Frank Hayes had recommended the hill as a likely place to drill as early as 1918, but management had ignored the recommendation.

In reopening the Signal Hill prospect, Shell assigned a young geologist named Dwight Thornburg to map the hill. Thornburg had grown up in Long Beach and remembered having seen marine fossils and tilted beds in a gravel quarry while playing there as a youngster. His surface geology reinforced the premise of Hayes that Signal Hill was the crest of a large anticline that conceivably could harbor a giant oil field.

At the time, Shell was riding a five-year wave of exploratory frustration and failure. The major area of interest was Ventura, which looked like a potentially big field, but up to then had defied efforts at commercial production. Shell had dropped $5 million in the five-year campaign and was seriously considering abandoning the prospect.

The company's California geologist, Alvin Schwennesen, proved a persuasive advocate for Signal Hill. His case was helped by the explosion of new cars on the nation's highways, creating a demand for gasoline that appeared to be only a small portent of what lay ahead. To ensure itself of a reasonable share of the growing market, Shell had to find and develop new sources of crude oil.

Landmen took to the field, spending $60,000 to lease 240 acres on Signal Hill, including a single large tract owned by Alamitos Land Company, which had not yet been subdivided. As workmen began preparing the location for Alamitos No. 1 near the intersection of Temple Avenue and Hill Street, the president of another large oil company was said to have remarked derisively that he would drink every drop of oil Shell found under the hill.

On March 23, 1921, a Shell rotary crew brought in from Coalinga spudded in to drill Alamitos No. 1. On May 2, the crew on tour pulled a core, recovering oil sand. The company appropriated $50,000 to lease more land. The crew ran casing to the bottom of the hole, and a cable tool crew arrived to finish the well. The rationale was that cable tools would avoid blocking off the oil sand with rotary mud. On May 23 while making a water shut-off test, the cable tool crew

December 1932 in the drafting room, Department of Petroleum and Gas, the Ferry Building, San Francisco. (Division of Oil and Gas.)

found 70 feet of oil standing in the hole, along with a substantial amount of gas. The news traveled fast, drawing spectators in such numbers that Shell finally had to erect a barricade to keep sightseers off the derrick floor. Drilling had reached a depth of 3,114 feet by June 23 when oil suddenly spurted over the crown block. The well sanded up and stopped flowing. Crews labored to clean out the hole.

When Alamitos No. 1 started flowing to tanks at four o'clock in the morning on June 25, an estimated five hundred spectators were on hand. During the first 24 hours, the discovery well produced 590 barrels of 22-gravity oil valued at $1.50 a barrel. Soon, production increased to 1,200 barrels per day, leaving no doubt that Shell's five-year dry spell had ended.

Flush production from closely-spaced wells led some to describe Signal Hill as "The World's Richest Oil Field." (Long Beach Public Library.)

Though only a few houses had been built on Signal Hill, most of the more desirable lots offered by subdividers had been sold. The patchwork of townlots formed the stage for one of the wildest land rushes California had ever seen.

The law of capture prevailed, with whoever could get a well down first likely to get the largest share. One approach simply was to buy property. The mansion that Andrew Pala had built as a retirement home on the crest of Signal Hill had been valued at $15,000 before the discovery well came in. Before

The "Spud Inn" restaurant at 20th and Obispo Streets on Signal Hill was open day and night. Bob and Butche's welding shop occupied quarters next to the restaurant. (Long Beach Public Library.)

midnight on the day the well was completed, Pala had received a cash offer of $150,000 for the house and lot. Shell succeeded in buying the property and turned the mansion into bachelor quarters for the stream of employees the company brought in to develop the field.

In the competition for leases, the royalty rate climbed from the one-eighth Shell had paid for the Alamitos Land Company acreage to one-sixth, and from there to one-third and finally to an unheard of fifty percent.

Speculators formed new companies overnight. To market their shares, promoters brought busloads of potential investors out from Los Angeles to tents they had set up on Signal Hill, where they provided the visitors with a free meal and a lecture by an "expert" extolling the easy money to be had in oil investments. The atmosphere of hustle and bustle, roaring engines and clanging pipe, flaring gas and escaping steam, and above all the redolent smell of crude oil, proved to be a contagious one for many who developed a severe case of oil fever, leaving an expectant investor to contemplate the riches he would receive from what might really amount to no more than one one-hundred-thousandth of a one-sixth royalty in an oil well that had not yet been drilled.

In the months that followed the discovery, drilling activity increased until Signal Hill bristled with derricks, some built so closely together that the derrick legs interlocked. On windless days the hill was shrouded in fog formed by steam from the boilers that drove the drilling rigs. The tight spacing of rigs produced some interesting situations. In one instance, a crew fishing for a string of tools reportedly snagged the fishtail bit from a neighbor's well. In another, a foreman was said to have berated a crew he accused of doing a sloppy job, only to find he was on a competitor's rig.

In the wake of the discovery, the Long Beach Women's Club held a public discussion to debate the pros and cons of turning the city's business district into an oil field. The city manager said if the wells

INCORPORATED UNDER THE LAWS OF THE STATE OF DELAWARE

JULIAN PETROLEUM CORPORATION

TOTAL AUTHORIZED CAPITAL STOCK OF 200,000 SHARES OF 8% CUMULATIVE PREFERRED STOCK OF THE PAR VALUE OF $50.00, AND 200,000 SHARES OF COMMON STOCK WITHOUT NOMINAL OR PAR VALUE

NUMBER — SHARES

PREFERRED STOCK $10,000,000 PAR VALUE $50.00 NON ASSESSABLE — COMMON STOCK NO PAR VALUE

THIS IS TO CERTIFY THAT is the owner of Shares of the par value of Fifty Dollars ($50.00) each of the Preferred Capital Stock of JULIAN PETROLEUM CORPORATION transferable only upon the books of said Corporation by the holder hereof in person or by duly authorized Attorney upon the surrender of this Certificate properly endorsed.

The preferences to which the Preferred shares of the capital stock of this corporation are entitled are set forth in full on the back hereof.

IN WITNESS WHEREOF, the said Corporation has caused this Certificate to be signed by its duly authorized officers and its Corporate Seal to be hereunto affixed this day of 19

PREFERRED CAPITAL STOCK

JULIAN PETROLEUM CORPORATION

SECRETARY — PRESIDENT

W P JEFFRIES COMPANY LOS ANGELES SPECIMEN

At Signal Hill, burning flares from the natural gas released in drilling operations fostered a carnival atmosphere in which promoters like the infamous C. C. Julian sold shares to a gullible public. (Long Beach Public Library.)

were big enough, it might pay to replace business blocks and residences with derricks and refineries. He cautioned, however, the city should not be in a hurry to tear down buildings until they had some assurance the wells would be big.

By the summer of 1923, 270 drilling rigs were working in the field, including 75 operated by Shell. In October of that year, production at Signal Hill peaked with an output of 259,000 barrels per day, representing one out of every three barrels of oil produced in California, which with an output of 812,000 barrels per day was the nation's top oil-producing state.

Through the Signal Hill boom, gas flares burning in the field often provided enough light to permit men to continue working at night without any other source of illumination. In October 1923, the Department of Petroleum and Gas reported that out of 290 million cubic feet per day of gas being produced with oil at Signal Hill, only 70 million cubic feet, or 24 percent, was being utilized for domestic and industrial purposes. The remaining 220 million cubic feet was being "wasted into the air."

At this time, the Department took an increasingly dim view of the gas wastage, whose fundamental cause was unrestrained reservoir development. In August 1923, State Oil and Gas Supervisor Collom and R. M. Barnes, the Coalinga deputy, wrote in the monthly bulletin, "...the great production of the Los Angeles Basin has placed the California petroleum industry closely upon the borders of disaster...The uncontrollable, in fact enforced, drilling of small property holdings has shown the futility of attempting conservation where each man is set against his neighbor in a greedy contest to get the lion's share of this mobile resource."

The discovery at Signal Hill, named the Long Beach field, was followed by the discovery of four more giant fields during the next three years in the Los Angeles Basin, including Torrance, Dominguez, Seal Beach and Inglewood.

The Department singled out Dominguez, a 1923 discovery by Union Oil Company, as an example of the correct way to go about developing a field. Acknowledging that the task was facilitated by the fact that virtually all of the productive acreage was controlled by three companies, Harold V. Dodd, District 1 petroleum engineer, described Dominguez development as being "along lines insuring the greatest economy and probably also the greatest recovery of oil and the least waste of gas."

It was said that American Indians had once stood on Signal Hill to signal their compatriots on Catalina Island, some 35 miles across the water, and that Spaniards had later built beacon fires on the promontory to guide ships at sea. The towering torches that accompanied blowouts in the 1920s sent a message of one of Southern California's biggest oil booms. (Los Angeles Public Library.)

An oil discovery in 1929 brought a drilling boom to Playa del Rey and Venice, which were 15 miles southwest of downtown Los Angeles. (C. C. Pierce Company, Seaver Center.)

The discovery at Signal Hill was followed by a series of new field discoveries before the end of the decade in the Los Angeles Basin, including Playa del Rey field in 1929. (Long Beach Public Library.)

Burning well fires that lit up the night sky would be an enduring memory for many who lived through the great oil booms of the 1920s in the Los Angeles Basin. (Long Beach Public Library.)

Under cooperative agreement, Union, Shell and Associated Oil Company uniformly spaced wells about 600 feet apart. And scarcely more than two years after the discovery, Shell and Union started gas injection activities that would prolong the flowing life of producing wells. Before the end of 1926, all of the approximately 36 million cubic feet per day of gas produced in the field was being utilized without any wastage, Dodd reported.

The Department, through its monthly bulletins, built a strong case for the conservation of gas with a series of articles by Department personnel and concerned industry representatives.

R. D. Bush, who succeeded R. E. Collom as State Oil and Gas Supervisor, wrote of the goal of recovering more of the oil in place, "Probably nothing now known will accomplish more to that end than the conservation of gas in the production of oil, particularly in fields of the future still undiscovered and in comparatively new fields where there are yet appreciable gas pressures." Commenting on fields where the wastage of gas figured to leave from 75 to 85 percent of the oil in the ground, Bush wrote, "More will probably be accomplished in increasing the percent of recovery by restoring pressure with gas or air forced into the oil sands through exhausted wells than by any other known method, except mining, which has a very limited application."

A. F. Bridge, vice president of Southern Counties Gas Company, wrote that the waste of gas was "aggravated by the 'get it now or never' drilling activity of a townlot field." He noted that in the Santa Fe Springs field gas production through wastage had declined from 510 million cubic feet per day to 50 million cubic feet in 11 months, and in the Alamitos Heights sector of the Long Beach field from a peak of 102 million cubic feet per day to 29 million cubic feet in only three months.

In another bulletin, the Department furnished a forum for the findings of a subcommittee of the California Operators General Committee on Gas Conservation, appointed by E. W. Clark, president of the American Petroleum Institute. The main conclusions were that the conservation of gas was vital to obtain the greatest ultimate recovery of oil with the least cost and that the best results could be obtained by the control of fields as a whole. The report advised, "Gas-oil ratio control not only pays dividends to the operator in the long run through greater recovery of oil but also conforms to the public demand for the conservation of gas."

Blowouts were another problem. At Signal Hill, Shell was drilling Nesa No. 1 on the crest of the hill a short distance east of Cherry Avenue. Rotary crews drilled top hole, ran casing as had been done in the discovery well and moved out so that cable-tool crews could carry the well on down through the oil sand. After cement for the casing had set up for two weeks, a cable-tool crew started drilling out the plug

Sometimes wooden derricks, though not immediately involved, were casualties. Above, deliberately "throwing" a derrick to prevent fire from spreading. (Long Beach Public Library.)

of cement that was left inside the well. When they penetrated the cement, gas that had collected while the cement was hardening blew out and caught fire quickly. The flaming torch could be seen from thirty miles out to sea. The roar could be heard for miles around.

For two days, the fire burned out of control while Shell's crews ringed the site with boilers. The flaming well and the preparations to control it drew a horde of sightseers. Finally, jets of mud and steam were directed at the well, and the fire was snuffed out, though gas continued to escape. At this point, weary firefighters looked up to see a spectator calmly smoking a cigarette only a few feet away. Needless to say, the cigarette was put out quickly. Eventually, the well was completed and became one of the field's best producers.

In the Santa Fe Springs field, Union Oil Company was at 2,060 feet at Alexander No. 1 when mud began to boil out of the casing. The driller shouted at the crew to run. A man up the derrick started to climb down. When he was about thirty-five feet from the ground, a column of mud and rocks shot out of the hole, rising to the crown block. The derrickman jumped and fortunately landed in the sump. Though shaken, he escaped with his life. Drill pipe blew out of the hole, which then cratered, swallowing up the drilling rig and the two automobiles the men had driven to work. Within hours, the site

Fighting a well fire at Santa Fe Springs, 1920. (Long Beach Public Library.)

The spectacle of a well afire quickly drew a crowd of onlookers. At one Signal Hill conflagration, workmen after a two-day battle succeeded in snuffing out the flames, but gas continued to escape. At this point, weary firefighters looked up to see a spectator calmly smoking a cigarette. (Los Angeles Public Library.)

of the well was a cauldron filled with swirling mud and debris.

In the same field, George F. Getty Inc. was deepening the SFS No. 17 to the Buckbee zone on the Nordstrom lease one Sunday, and lost circulation while coring oil sand at 5,409 feet. As the crew was attempting to pump cement down the hole to regain circulation, the well blew in, flowing oil and gas for about 20 minutes before the gas caught fire from the boilers. The wooden derrick collapsed in minutes. The kelly fell out across Telegraph Road. As the kelly was attached to the drill pipe in the hole, it acted as a jet, shooting a stream of fire across the road. Two wooden derricks adjacent to the well burned down. Steel derricks close to the flames crumpled under the extreme heat. Fire also destroyed a small restaurant at Telegraph and Shoemaker Road and damaged a newly opened service station

In the wreckage of twisted metal and fallen timbers, workmen took shelter behind corrugated iron shields and in a small metal shed to direct streams of water against a flaming well. Rohde & McKeon lease, June 11, 1929, Santa Fe Springs. (California Oil and Gas Association.)

at the same intersection.

All the operators in the area tried to smother the well with mud and steam. When the effort failed, engineers from General Petroleum Corporation, whose lease offset the Getty property, took charge of the battle to control the well, with Getty agreeing to pay the expenses. At a distance of 200 feet from the blazing well, workmen erected an asbestos shield 100 feet wide and 20 feet high. Behind the shield, a steam shovel excavated a pit 20 feet deep, banking dirt around the edge to guard against the danger of a flood of oil.

The company brought in a mining engineer to direct the task of digging a 200-foot tunnel to the wellhead. The four-by-six-foot tunnel was timbered carefully as workmen drove the shaft toward the well. A track was laid to keep pace with progress, allowing dump cars to carry cuttings out of the tunnel. Three shifts of workmen labored around the clock. A conductor pipe for proper ventilation was carried forward. Chemists tested the air continuously to safeguard the men burrowing toward the well. As the number of men working on the project rose, the company's cost of insuring the workmen, based on a rate of $175 per day per man, soared to over $16,000 a day.

One week after the tunneling work began, men had reached some 190 feet in the shaft and were closing in on the well casing when they encountered quicksand. They raised the elevation of the tunnel two feet to get above it.

When the men were within a few feet of the casing, they drilled small-diameter coreholes to test the ground and detect and guard against any escaping gas. Danger mounted as they cautiously stripped dirt and cement from around the casing. Any break in the casing would mean almost certain death with the release of burning oil and gas into the tunnel.

The workmen built a chamber some 20 feet square and 9 feet high around the casing. They placed clamps on the 15-inch casing, then cemented the space between the 15-inch and the inner 11-inch string of pipe and stripped the 15-inch pipe with air cutters. They then stripped the cement from the 11-inch casing, placed additional clamps on top and bottom and carefully made contact with the 9-inch

Drawn by rumors of big pay, sturdy men flocked in from all parts of the country to join the Los Angeles Basin drilling boom. One veteran driller described his crew as composed of "one grocery clerk, one dry goods clerk, one cop and one college student." Above, Superior Oil Company's crew on Steinhilber No. 2 in the Torrance field, 1923. From left to right, they are Horne, Roneck, W. Fuqua, T. Wilson and J. Sullivan. (John C. Sullivan.)

casing string inside. They managed to tap the pipe and tried to construct a bridge to control the flow of oil and gas by inserting aluminum balls, lead wool "sausages," steel wool, bundles of hemp and asbestos rope. The effort failed.

They laid lead lines and succeeded in tapping the casing to carry off most of the production through the tunnel. With the fire still burning at the surface, about 3,000 barrels per day of oil and 7 million cubic feet per day of gas were recovered. When the fire at the surface finally was controlled with steam and mud a month after the blowout began, the connections were repaired and the well went on production making 5,500 barrels per day, cutting only 0.1 percent water, and about 9 million cubic feet per day of gas.

The State of California, concerned about the wastage of gas and the occurrence of blowouts, addressed these matters in August 1929 with amendments to the oil law.

The "Gas Act" was designed to control and prevent gas wastage. It stated that "...the unreasonable waste of natural gas...whether before or after the removal of gasoline from such natural gas, is hereby declared to be opposed to the public interest and is hereby prohibited. The blowing...of natural gas into the air shall be prima facie evidence of unreasonable waste."

The act provided two methods by which unreasonable waste of gas might be restrained. One involved an order by the State Oil and Gas Supervisor if, in a hearing, he determined there was unreasonable waste. If the order was not complied with, the Director of the Department of Natural Resources, which that same year superseded the State Mining Bureau as the parent agency, was authorized to institute proceedings in Superior Court. Or, if the Director perceived wastage, he was authorized to institute proceedings directly in the name of the people of the State of California.

The blowout prevention amendment required the installation of blowout prevention equipment on all wells and the inspection of the equipment by the Division of Oil and Gas, which had superseded the Department of Petroleum and Gas. Under the amendment, before drilling was begun below surface casing, a gate had to be installed for closing in the well with the drill pipe out of the hole, and a blowout preventer installed to close around the drill pipe once it was in the hole. Controls were to be located outside the derrick, a mud line with high-pressure fittings was to be connected into the surface casing below the blowout preventer, and there was to be a high-pressure stopcock on the kelly.

An inevitable result of the big discoveries of the 1920s was an over-supply of oil.

For Shell, increasing production from Signal Hill posed a problem. The company engaged tankers to carry the crude north to the Martinez refinery,

Accidents were part of a sad picture, the casualty statistics of work in California's oil fields. As the state's production rose to more than 100 million barrels a year, the highest it had ever been, the accident rate rose, as well. The Bureau of Mines tabulated a total of 4,109 accidents during 1921-1922 in the producing departments of ten representative oil companies operating in California. The state's Industrial Accident Commission listed 98 fatalities and 406 men permanently injured. (Los Angeles Public Library.)

which had been built to handle crude from Coalinga. Faced with an acute storage problem, Shell shut in 300 wells at Coalinga capable of producing 16,000 barrels per day. The company's engineers reversed the flow in the Coalinga-to-Martinez pipeline to use the 1,100,000-barrel tank farm at Coalinga and the 1,500,000 barrels of steel tankage along the pipeline for storing Signal Hill crude. A 750,000-barrel concrete reservoir that had been built at Coalinga before Shell's advent was reconditioned and put into service for additional storage.

Not all of the Los Angeles Basin discoveries contributed to the oversupply. When San Clemente Oil Company moved out 12 miles northwest of Long Beach to prove up the Lawndale field, speculators with visions of overnight riches promptly jumped on the bandwagon to make the new field another townlot bonanza. In summing up the results one and one-half years later, E. Huguenin, deputy supervisor for District 1, described the field as "probably the greatest failure in the history of California drilling operations." Of 63 wells drilled by various operators only 6 produced oil.

There were five other giant discoveries before the end of the decade. The first four included Shell's discovery of the Mount Poso field in Kern County in July 1926; Pan American Petroleum Company's discovery of the Rincon field in Ventura County in December 1927; General Petroleum Corporation's discovery of the Edison field in Kern County in July 1928; and Barnsdall Oil Company's discovery of the Elwood field in Santa Barbara County in July 1928.

The fifth and largest find came before the end of 1928 when Milham Exploration Company's Elliott No. 1 blew in at Kettleman Hills in November, flowing at a rate estimated at some 4,000 barrels per day. *California Oil World*, in its November 8, 1928 issue, described the 60-gravity oil as "almost unbelievably light in texture—so light, in fact, that trucks are filling up their tanks and using it to fuel their motors."

In a preliminary report on the Kettleman Hills discovery, E. H. Musser, deputy supervisor for District 5, said the new field "promises to be the greatest oil field of California."

The rush to develop Kettleman Hills came against the background of a prosperous country. Though oil was in oversupply, nationally the economy seemed in solid shape, with more news emphasis on the problems of enforcing prohibition and the attendant gang wars including a St. Valentine's Day massacre in Chicago in which seven men were killed, than with economic ills.

One index of prosperity was the report by the Automobile Club of Southern California on the phenomenon of the two-car family. More than three million American families, the Club said, now owned two cars, and the number was steadily growing. Various reasons were advanced for the growth of the two-car

Wilshire and Fairfax Boulevards, circa 1920. (Seaver Center.)

Wilshire Boulevard was little more than a country road when wildcatters began a determined search for new fields in the Los Angeles Basin. (Seaver Center.)

family, including such things as memberships in golf clubs, children at college, the increasing number of women drivers and the use of automobiles in business.

Production surged, and demand failed to keep up until the amount of available oil far exceeded the amount the market could absorb. By late 1929, California was producing 801,120 barrels per day, up some 524,000 barrels per day since the decade began, but with an additional 190,985 barrels per day of production shut in.

On the last Friday in October 1929, the Associated Press reported that Washington was keeping an eye on Wall Street, but though securities prices had slumped, the unofficial opinion was that the drop need have no depressing effect upon the nation's general business structure. On the following Tuesday, October 29, 1929, desperate speculators sold 16,400,000 shares of stock.

Shielded derricks protected a cemetery from unwanted sprays of oil. (Ansel Adams, Los Angeles Public Library.)

5 The Great Depression

Though the oil industry was severely crippled by the Great Depression, there was still time for an occasional showing of new equipment designed to advance the state-of-the-art in the oil fields. The model is standing by a Shaffer cellar control gate. (Los Angeles Public Library.)

At Mascot No. 1 on 25 Hill in back of Taft, Standard Oil Company early in September 1929 called out Deputy Supervisor E. H. Musser of the Division's Taft office to witness the water shut-off test for a string of 9-inch casing cemented at 6,350 feet.

The exploratory well had suffered frequent lost-circulation problems since the company spudded in eight months before to search for deeper oil that might regain for the Midway-Sunset field the leading role in the production of California's oil.

Because of developments in the Los Angeles Basin, the West Side field had been relegated to third place among California's fields after a reign of more than ten years as the most productive field. Thanks to deeper pool discoveries, Long Beach was now the top field, putting out 190,000 barrels per day, followed by Santa Fe Springs, 92,500 barrels per day, and Midway-Sunset, with 75,000 barrels per day.

Even as misfortune was stalking the East's financial establishment, Standard's crews were bailing fluid to a depth of 2,953 feet and allowing the Mascot well to stand for 20 hours, during which time 6-1/2 feet of fluid entered the hole, equivalent to one-half barrel per day. The water shut-off was approved.

Drilling resumed, with crews happy to put behind them the battles with lost circulation. In one instance, they had spent two days pumping in Aqua Gel—a mixture of silica and aluminum—mixed with clay-like Mojave mud, six bales of hay, 258 sacks of sawdust and more than a ton of cottonseed hulls. In another instance, they had regained circulation only after laboring four days to pump in 156 tons of dry Mojave mud mixed to a weight of about 80 pounds per cubic foot when commingled with 800 barrels of mixed mud, 290 sacks of cottonseed hulls, 17 bales of straw, 410 sacks of sawdust and 35,400 pounds of Aqua Gel. With pipe in the hole, drilling progressed with occasional shows of gas and oil in the ditch.

There seemed no immediate reason to believe there would be a depression in the oil fields. One day after "Black Tuesday" in late October, the Associated Press reported the stock market was returning to normal with scores of stocks on the New York Stock Exchange up $5 to nearly $30 a share. In Washington, D.C., R. Julius Klein, Assistant Secretary of Commerce, went on the radio to say that American business need expect no adverse results from the collapse of stock prices. Klein said less than one percent of the population had been affected by what he described as the "speculation gyrations."

On the following day, K. R. Kingsbury, president of Standard Oil Company, issued a reassuring statement. "I know of nothing in the present condition or the future of the oil industry that justifies the extraordinary depreciation in prices of oil stocks today," he said. "No new conditions have arisen threatening our company or, so far as I know, the industry. On the contrary, the prospect for effective conservation, which will result in balancing production with demand, is brighter today than it has

Shell Oil Company runs the first electric log in the United States in 1929, near Bakersfield.

been at any time."

Stock prices continued to decline. President Hoover announced a plan to seek stabilization by tapping the $250 million fund established by the Jones-White Act to assist in the expansion of the Merchant Marine. Shipyards would be booming in six months, newspapers reported. A week later, representatives of the nation's public utilities announced plans to spend $1.5 billion in improving and expanding their systems during the coming year. A few days later, Ford Motor Company announced a wage raise totaling $20 million, boosting the minimum wage for its employees from $6 to $7 a day. Nothing turned the economy around. By the end of the year, the stock market crash had cost investors an estimated $40 billion.

On 25 Hill, Standard crews drilled ahead. At 7,000 feet, they cut through shale fractures that showed gas and light oil. Below 7,800 feet, they ground away at hard silty sands and shales, coring almost as much as they drilled, getting some shows. By Christmas, the well was below 8,000 feet.

From Southern California came word of a new world depth record. Oscar Howard's Hathaway No. 7 near the corner of Telegraph and Norwalk Roads in Santa Fe Springs had gone to 9,350 feet and was testing.

Three days later, the newspaper carried a short notice of another economic disaster in the East. Stutz Motor Company, whose products had long been admired in the oil fields, was going into bankruptcy.

While Standard continued to make hole, there were increasing signs that the oil fields might not be immune to the economic problems afflicting the rest of the country. There began to be a hollow ring to "blue sky" pronouncements of good times coming when they were measured against a sharp drop in building permits, refineries going on a six-day week to hold down product inventories and talk of a six-hour day on drilling rigs to keep a maximum number of men working, even if each man earned less. For the operator, the arrangement was said to offer one notable advantage: men would not have to be given time off to eat as on eight-hour tours.

At the Mascot well, crews passed 9,200 feet, coring continuously and making about fifteen feet a day. Twelve hours or more were required for each round trip to pull the core from the hole and go back in to cut another. In late March 1930, the crew cut a core taking the well to 9,629 feet and making Standard's Mascot No.

Standard Oil Company's Mascot No. 1 on 25 Hill in back of Taft was the world's deepest well, but the technological achievement was clouded by the Great Depression. (Petroleum World.)

1 the deepest well in the world by almost 300 feet. Only one foot was recovered from the 18-foot interval cored. A. D. Henderson Jr. described the core as "very hard poorly sorted medium grained gray sand, faint odor light oil, no cut with carbon tetrachloride, fair cut with acetone."

Of the depth record, *California Oil World* noted, "Geologic data, some of which may prove of the utmost scientific value, constitutes practically the only value of the deep drilling Standard is carrying on in the Mascot well...even if oil were found now, it would require a big well with ample gas pressure to flow it to make it of commercial value." However, the writer was not entirely pessimistic, and noted, "This well being the first to tap the earth to this depth, everything that its log reveals will be of importance. The Mascot may develop information of great value not only to the geologist and to the scientific world but also to the driller and to the manufacturer of drilling equipment." The writer concluded, "When oil is higher in price, even as deep a well as this might pay for operation."

In April, crews ran the longest string of casing ever run, cementing 5-3/4 inch casing to a depth of 9,302 feet. They installed a Shaffer-Warner casing head and closed in the well. Standard Oil Company announced operations would be suspended until July 1931, more than a year away, or later, in compliance with a plan by the company and other majors to hold down activity in California fields "to eliminate the over-production which has featured the industry for some months."

At the time, the community of Taft was within two months of having its first marathon dance at Buchanan's Pavilion. Seven couples would compete for $1,000 in prizes in an attempt to break the world's marathon dance record of 1,685 hours. None would break the record, though two couples would succeed in dancing forty minutes out of each hour, resting twenty minutes, for more than 280 successive hours, or almost twelve straight days.

Before the year ended, more than six million Americans were out of work. The number rose to twelve million during the following year. While Mascot No. 1 stood shut in, the country sank steadily deeper into the worst depression in its history. More than 5,000 banks failed and over 32,000 businesses went bankrupt. Desperate men sold apples on street

M. D. Freeland was one of the Standard Oil Company drillers who helped make Mascot No. 1 a record-breaker. (Petroleum World.)

corners, ate in soup kitchens and lived in collections of shacks like the one of cardboard, corrugated iron and scrap lumber that took shape at the eastern end of Center Street in Taft.

Though times were hard and would get harder, the oil industry did not come to a complete standstill, nor did the Division of Oil and Gas cease to push for the reforms that would make the industry a stronger, more acceptable partner in California's economic development.

One ongoing activity involved enforcement of the Gas Act that had been passed in 1929. The act had scarcely become law before the first action was filed against operators in the Santa Fe Springs field, where an estimated 500 million cubic feet per day of gas was being blown to the air. The Superior Court issued a preliminary injunction that reduced wastage by more than 100 million cubic feet per day. During the legal sparring that followed, natural decline further cut wastage and the case was dismissed two and one-half years after it had been initiated.

Blowouts and the destructive fires that often accompanied them proved a continuing concern in the Los Angeles Basin. (Los Angeles Public Library.)

In the Ventura field, an estimated 70 to 135 million cubic feet per day of gas was being blown to the air. Since only six operators were involved, administrative hearings were held and wastage was reduced.

The next target was the Long Beach field, where 100 million cubic feet of gas per day was being wasted. The Superior Court set a trial date, but the case was dropped as wastage was reduced, largely by natural decline.

At Kettleman Hills, the Division filed action to reduce the wastage of 180 million cubic feet of gas per day. The court issued a preliminary injunction, shutting in wells with high gas-oil ratios with the goal of reducing wastage to a 10 percent limit. Some three years after the suit was filed, wastage was down to 3.7 percent. The Supreme Court of California later ordered the Superior Court to dismiss the case. At the time, wastage was about 1.7 percent. No new action was initiated.

Before the decade of the 1930s ended, the Division through its legal actions had played a major role

One of the targets for enforcement of the Gas Act that had been passed in 1929 was the Santa Fe Springs field, where an estimated 500 million cubic feet per day of gas was being blown to the air. (Long Beach Public Library.)

in reducing wastage in the Elwood field from 22 million cubic feet per day to less than 2 million cubic feet, in the Dominguez field from an estimated 20 million cubic feet per day to scarcely more than one million cubic feet, in the Playa del Rey field from about 30 million cubic feet per day to less than 3 million cubic feet, in the Mountain View field from about 40 million cubic feet per day to only slightly more than 3 million cubic feet and in the West area of the Montebello field from 30 million cubic feet per day to less than 6 million cubic feet.

Though progress was being made, the battle was not yet over. In July 1938, Petroleum Securities Company brought in a 4,776-barrel-per-day discovery well at East Coalinga Extension field, seven miles northeast of Coalinga.

Development proceeded rapidly with the drilling of 157 wells during the next three years. Early in the life of the field, a gas cap began to form at the crest of the structure. In an effort to conserve gas energy, operators formed a unit in March 1950 that covered over 94 percent of the approximately 3,200 productive acres.

Los Nietos Company, unit operator, began gas injection on a small scale in April of that year. Full-scale injection began in December, but had to be discontinued by the end of the month because of high-volume gas production on a nonunit lease. In May 1951, members of the unit acquired the lease.

Full-scale gas injection resumed. Gas soon broke through into two wells beyond the boundary of the unit. By December 1951, the gas production from the two wells had increased from the 1950 average of about 0.4 percent of total field gas production to 8.6 percent. By July 1952, the percentage had increased to more than 20 percent.

Meetings were held in the latter part of 1952 and early part of 1953 in an effort to solve the problem voluntarily. When it became clear such a solution could not be achieved, DeWitt Nelson, Director of Natural Resources, authorized legal action. A complaint filed on July 16, 1953, in the Superior Court of Fresno County alleged unreasonable waste of gas by ten unit and nonunit operators. State experts predicted that a probable result of wasteful production practices would be the loss of some 100 million barrels of oil.

After a hearing in which demurrers were overruled and several motions denied, the preliminary injunction requested by the state was granted on July 30, 1953. Under the injunction, a three-member committee was formed to administer the terms. The members included one named by the State Oil and Gas Supervisor, one by nonunit operators and one by the unit operator. The injunction provided for a testing period during August 1953, with a complete shutdown from August 2 to August 17 for the measurement of bottomhole pressures. Following completion of testing, the committee allocated monthly oil production on the basis of gas-oil ratios.

Immediately before the injunction, the total field gas-oil ratio was over 1,000 cubic feet per barrel, with an average of 528 for unit-operated wells, and approximately 17,500 from three nonunit wells. Six years later, the field average was about 560. Oil production in that same period declined from about 45,900 barrels per day to 38,300 barrels, a drop of 16.5 percent, or 2.75 percent per year average. Operation of the field under the injunction clearly was a success from an engineering and conservation standpoint.

Gas wastage was not the only problem that concerned the Division of Oil and Gas during the 1930s. In the matter of continuing reform, the Division in 1931 helped secure the passage of legislation dealing with bonding and spacing requirements.

As a reform of bonding requirements, legislation was passed in 1931 requiring the posting of a $5,000 bond for drilling, redrilling or deepening all wells. The bond would be forfeited in the event an operator failed to properly abandon or otherwise operate the well. The forfeited money would provide the Division funds to carry out the work.

The spacing statute was a response to haphazard townlot drilling. The statute restricted drilling to parcels no less than one acre in size. In addition, wells on leaseholds greater than 250 feet in width had to be located 150 feet apart and 100 feet from property or street lines. On parcels that were less than 250 feet in width, the well was to be located as far from the lateral boundaries as possible with due consideration for existing improvements. Engineers hailed spacing provisions as ultimately resulting in the recovery of more oil and a longer life for producing fields.

Almost half a century later, another milestone would be reached in the matter of well spacing. In September 1977, Occidental Petroleum Corporation discovered the Cal Canal field, 35 miles west of Bakersfield. As development began, the company filed a petition with the Division of Oil and Gas asking that no more than one well per 40 acres be permitted for the Stevens zone, which was productive at an average depth of 11,800 feet.

After a public hearing in 1978, M. G. Mefferd, State Oil and Gas Supervisor, approved the 40-acre spacing plan. One of the affected parties, Getty Oil Company, which held acreage offsetting production developed by Oxy, appealed the Supervisor's order. The appeal was denied by Priscilla C. Grew, Director of Conservation.

The spacing decision was precedent-setting. For the first time in the history of the Division of Oil and Gas, spacing other than the basic one-acre was set by the State Oil and Gas Supervisor under authority granted in 1974 by the Public Resources Code. The Cal Canal order also required that a pooling agreement be established in a specified portion of the field to protect correlative rights. Although the Supervisor had the authority to order mandatory pooling, in this instance the parties involved—Oxy and Getty—reached

At Signal Hill, some claimed it was possible to walk great distances from rig to rig without ever touching the ground. The Division of Oil and Gas helped secure passage of spacing legislation hailed by engineers as a significant step toward recovery of more oil and longer life for producing fields. (Long Beach Public Library.)

a voluntary agreement.

As the Depression continued, the number of new well notices filed with the Division declined, slowly at first, then precipitously. The number of notices dropped from 1,256 in 1929 to 918 in 1930, and then on a roller-coaster ride to 329 in 1931 and a low of 279 in 1932, representing a decline from pre-crash days of 78 percent in only three years.

A few wildcatters continued to look for oil. In September 1931, a company calling itself Ranger Petroleum Corporation incorporated with $75,000 authorized capitalization. From an office in the Pacific Southwest Building in Long Beach, the company directed the drilling of a well designated Watson No. 2 at Wilmington. In January 1932, the company completed the well on a pump for 150 barrels per day of clean 14-gravity oil from a 100-foot interval at 3,674 to 3,784 feet. Ranger found a limited market for the crude, selling it as fuel for the Dollar Lines through the Olympic Corporation.

The discovery did not create any excitement because of the low gravity of the oil and the modest production. The assumption was that it was merely an extension of the Torrance field, which had been discovered ten years before. In addition, there was no mention of the discovery well in the *17th Annual Report of the State Oil and Gas Supervisor*.

It would be four and one-half years later before the Division would designate Wilmington as a new field. The designation followed the completion in June 1936 of a Harbor Drilling Company well at Mauretania and Coil Streets in the Ranger zone, pumping 130 barrels per day of clean 16.8-gravity oil. The well found the top of the pay zone some 200 feet higher than in the discovery well and this, along with the higher gravity of the oil, was taken as proof that Wilmington was a new field.

In December of that year, General Petroleum Corporation completed Terminal No. 1 in the Los Angeles Harbor area for a flow of 1,350 barrels per day of clean 20.5-gravity oil. The well proved up three zones that had not been tapped previously, showing Wilmington was much larger than originally indicated. Ironically, the field discovered in the depths of the Great Depression would prove to be the largest oil field in California.

Five months after Ranger Petroleum made the unheralded Wilmington find, Standard Oil Company crews began rigging up at Mascot No. 1, which had been suspended for more than two years. While the well had been shut in, the world's depth record had passed out of the country to the state of Veracruz, Mexico, to Penn-Mex Fuel Company's Jardin No. 35, which had gone to 10,585 feet.

On July 6, 1932, Standard's crews began the long road back at Mascot No. 1. By late October, they had gone to 9,753 feet and were trying to get back on bottom with a 4-3/4 inch bit when the bit was blocked by an obstruction. Efforts to clean it out failed. They could get pipe in the hole, but could work up only

inches at a time and then with maximum allowable strain and great difficulty. They could go no deeper. With difficulty, they cleaned out to 9,710 feet and ran in with a Johnston formation tester to test on bottom. Only a small amount of gas flowed.

Continuing to test, they slowly came back up the hole, getting no encouragement. In March 1933, they began cutting and pulling casing to salvage what they could from the unsuccessful well that once had been the deepest in the world, but with its abandonment marked the end of one more dream in the hard times of the 1930s.

Some six months after Standard called it quits at the Mascot well, the Federal Government in September 1933 effectively took control of production rates. For California, federal control ended more than three years of voluntary curtailment. It meant tightening crude curtailment strings, cutting refinery runs to market requirements and new contracts for millions of barrels of heavy fuel oil for the Atlantic seaboard as a direct result of the further curtailment of South American imports.

At the time the Federal Government took control, production in California had sunk to a post-crash low of 471,600 barrels per day. In the following year, indexes began to move upward again. New well notices climbed from the low of 279 in 1932 to more than double that number in 1933, when 596 notices were filed with the Division. Production was slower to respond, but by 1934, it had climbed some 5,000 barrels per day and would continue the upward climb to a high of 683,300 barrels per day in 1938.

As if to underscore the recovery that was beginning to stir in California, North Kettleman Oil & Gas Company drilled a 10,944-foot well at the northwestern end of the Kettleman Hills field, returning the world's depth record to the state. In June 1934, General Petroleum Corporation broke that record with Berry No. 1 in the South Belridge field, which bottomed at 11,377 feet.

An editorial in *California Oil World* commented, "Twenty years ago a 2,000-foot oil well was considered deep, 4,000 feet the limit. Ten years back we thought 6,000 feet was almost beyond the practical, though sometimes reached. We now theorize about going 15,000 feet sometime in future years. Whoever tries this first may fail, for no one knows what lies so deep, but later somebody will succeed and

From an unheralded discovery by a wildcatter in the depths of the Great Depression, the Wilmington field emerged as California's largest oil field. (Spence Air Photos, Long Beach Public Library.)

greatly increase our geological knowledge whether or not he gets oil."

In Northern California, wildcatters searched for new fields. Instead of oil, they found natural gas. In August 1935, Amerada Petroleum Corporation discovered the first commercial gas field in Northern California one mile north of Tracy. In early June 1936, Standard Oil Company completed the discovery well for the McDonald Island gas field. A little more than two weeks later, Amerada brought in the discovery well for the Rio Vista gas field, which would prove to be the largest gas field in California. The development of the fields assured the San Francisco Bay industrial area of a valuable gas reserve within a reasonably short distance and started the Sacramento Valley on the path toward becoming California's foremost gas province.

In the San Joaquin Valley, exploration moved away from surface structures to the flat floor of the valley, using seismic prospecting to find subsurface structures. One of the new tools was the reflection seismograph. The approach involved drilling a hole several inches in diameter to a depth ranging from a few feet to several hundred feet, sufficient to penetrate below the layer of loose surface material. An explosive charge was placed at the bottom of the hole, with seismic detectors positioned at various distances from it. When the charge was fired, the shock waves reflected off the different rock beds below were picked up by the seismic detectors and recorded in an instrument truck. The procedure would be repeated at each location in a previously determined pattern. The depth of a particular rock stratum would be determined at each detector station by the length of time required for the shock waves to return to the surface. Seismic profiles could be made in several directions across the area, and from this record the depth and dip of beds could be determined and the type of structure deduced.

Shell Oil Company was one of three companies to shoot an area in the Kern delta southwest of Bakersfield with a reflection seismograph, but only Shell seemed to perceive a structure. The company leased 6,433 acres from Kern County Land Company, which years earlier had fenced in a ten-section tract in the area, giving rise to the name Ten Section.

On the basis of information gained from the seismic survey, Shell spudded in to drill Stevens A No. 1 in the winter of 1936, launching the venture in the face of widespread opinion that the floor of the valley would prove to be a wildcatter's graveyard. On June 2, 1936, the Shell wildcat tested at a rate of 1,200 barrels per day of 60.4-gravity condensate and 15 to 18 million cubic feet per day of gas from a total

Four and one-half years after the depression-era wildcatter found California's largest oil field, another proved up the state's largest gas field. Above, Amerada Petroleum Corporation's discovery well at Rio Vista. (Krug Dunbar.)

depth of 7,880 feet. The discovery not only opened up a new oil field but also gave a name to what would become one of Kern County's most prolific oil sands—named by Shell the Stevens sand. The name was chosen because, near the discovery well, the Southern Pacific Railroad some years before had laid out a subdivision—still not settled—and called it Stevens Siding.

The Ten Section discovery well hardly had been turned to the tanks before other reflection parties spread out to shoot seismic surveys of the flat floor of the San Joaquin Valley. Before the year was out, Standard Oil Company had proved up a giant field on Kern County Land Company property at Greeley, six miles north of the Ten Section field.

From Greeley, the search moved farther to the northwest to a surface high that had looked interesting as early as the summer of 1924 when Union Oil Company had drilled an unsuccessful wildcat to a depth of 6,053 feet to test the structure. The advent of the reflection seismograph gave the company another tool with which to appraise the structure. A seismic survey indicated a subsurface high about one mile east of the topographic high. The company moved in a rig to drill an 8,500-foot wildcat to look at the Stevens sand that had yielded discoveries at Ten Section and Greeley.

In late March 1937, the company spudded in to drill Kernco No. 1-34 on a Kern County Land Company lease. Sixty-nine days later, the well was in Stevens sand. The sand yielded only high-pressure salt water and a little gas. The company decided to go deeper. At 10,200 feet, the wildcat was still without pay sand and rapidly approaching a depth beyond which no one had ever found oil.

Company engineers argued it was time to give up the Rio Bravo venture, declaring that such deep wells would not pay drilling costs, even if oil were found. Some said any sand found at that depth would be so tight from the weight of earth resting on it that it could not produce anything.

However, another sand in the area had been found productive in fields on the east side of the San Joaquin Valley where sands were much shallower than on the floor of the valley. It was the Vedder sand, which had been proved up by Shell with the discovery of the Mount Poso field in 1926. Those arguing for continuing on to the Vedder at Rio Bravo won, and drilling resumed.

In October, crews cored an interval from 11,236 feet to 11,255 feet and recovered oil sand. For fifty-five feet, they cored gradually darkening sand. They bottomed at 11,302 feet and ran an aluminum liner, figuring that if they decided to go deeper they could

From a search for surface structures, geologists turned to new tools with which to identify subsurface structures. On the flat floor of the San Joaquin Valley, Shell Oil Company discovered the Ten Section field with the Stevens A No. 1, at right. (Ed Burtchaell.)

easily drill out the liner. They set the pipe twenty feet above the sand formation, completing the well "barefoot."

The well came in flowing at the rate of 30,000 barrels per day for four and one-half hours before all available storage was filled. When cut back to a 40/64-inch bean, a restrictive device designed to slow the rate of flow, the well produced 2,000 barrels per day of high gravity oil. At 11,302 feet, Union's discovery well not only was the first well in California to produce oil from below 11,000 feet, but also was the deepest producing well in the world.

Within the next two years, six more new-field discoveries followed on Kern County Land Company acreage within a 20-mile radius of the initial Ten Section find, including the giant North Coles Levee field and the smaller Canal, Canfield Ranch, South Coles Levee, Strand and Wasco fields.

While reserves proved up at Wasco were not as great as the other fields, the field provided a laboratory for deep drilling. In April 1938, Continental Oil Company bottomed the 3,000 barrel-per-day discovery well at 15,004 feet, making the well the first in the world to go below 15,000 feet. The last bottom-hole temperature recorded was taken at about 13,000 feet. The reading was 284 degrees Fahrenheit, which was high enough to completely destroy the batteries of the surveying

The discovery of the Ten Section field in 1936 touched off a series of new-field finds in Kern County based on seismic prospecting. Shell Oil Company recognized the Ten Section discovery with a sign on the discovery well, which had been redesignated KCL-A No. 1-29. (Kern County Land Company.)

Commemorating the 30th anniversary of the Ten Section discovery, H. L. Reid, Kern County Land Company's vice president-oil and minerals, at a meeting in Bakersfield pointed out the location on the Kern delta map to two former Shell Oil Company associates who had worked on the historic well, Claude E. Peavy, left, former division production superintendent for Shell, and Dr. E. Fred Davis, right, former vice president of Shell in charge of exploration. (Kern County Land Company.)

instrument even though the instrument was encased in dry ice when the run was started. Though the field would produce only five million barrels before its abandonment in 1960, it would serve as a valuable proving ground for the development of deep drilling tools and techniques.

Though the giant fields discovered during the decade of the 1930s—from Wilmington on north to Coalinga and the huge East Coalinga Extension field—had proved up reserves of some four billion barrels, California's production during the decade had been surpassed by Texas' spectacular 4.01 billion barrels. But, California's production of 2.08 billion barrels, though second to Texas, still represented 20.3 percent of the nation's 10.2 billion barrel-production.

As the decade ended, both California and the nation faced a threatening future in a world that had gone to war again.

Union Oil Company's Rio Bravo discovery well proved to be the deepest producing well in the world. Above during the well test, the discovery well is illuminated temporarily by flares of gas burned to eliminate any possible hazard from gas accumulations. (Pacific Oil World.)

6 The War Effort

As the war called away younger men, older rig hands like derrickman Guy Miller, a Navy veteran of the first World War, doubled over on tours for Standard Oil Company's Northern Division, drilling new wells in the southern San Joaquin Valley. (Chevron Corporation.)

Margaret Binkley was living in the Dutch East Indies with her husband, Bill, an employee of Royal Dutch Shell, when the Germans invaded Holland in 1940. The Binkleys had met and married while both were students at the University of California in Berkeley. They had graduated in 1938, and Margaret had earned a degree in geology and paleontology.

The posting to the Dutch East Indies that followed Bill's employment by Shell had been the latest chapter in an ongoing series of moves for Margaret Binkley. Born in Inspiration, Arizona, she was the daughter of a safety engineer whose work took him to construction, oil and mining camps in the United States and foreign countries. She had attended her first school in Chile, where her father was working for a construction firm. She spent her freshman year in high school in Peru. While she was growing up, her father had let her ride with him while he patiently told her what was going on. He had a fine collection of ores and minerals, which fascinated her. She had always liked to know the what, how and where of things. The science and engineering fields provided tools with which to discover these answers.

On December 7, 1941, when the Japanese bombed Pearl Harbor, the Binkleys were living in California with their then one year old daughter, Mary. With the entry of the United States into the war, Bill Binkley volunteered for the Army Air Corps, which sent him to the Southwest Pacific. Margaret went to work for Standard Oil Company as a junior geologist in Taft, until a siege of valley fever took her off the job market. After recovering, she began looking for another job in Taft, where she had rented a house. The Binkley family had grown to two children now, and both were with her.

When she heard the Division of Oil and Gas was looking for an inspector for the Taft office, Margaret applied, and after an interview was told she could have the job if she could pass an examination. With the war on, she surmised they were so desperate they would have hired a green gorilla if one had passed the test.

While her degree was in geology and paleontology, she had taken some courses in petroleum engineering. She passed the state examination and became the Division's first woman petroleum geologist.

As Margaret Binkley's life was being reshaped by the war, the California oil industry, too, was undergoing profound changes.

One change had been taking shape even before the nation's entry into the war. In May 1941, the drift of events in Europe led President Roosevelt to declare an unlimited national emergency and to create by executive order the Petroleum Administration for Defense, headed by Interior Secretary Harold L. Ickes. Ickes had named as his deputy Ralph K. Davies, sending a clear message to the oil industry that the government was in earnest about seeking its cooperation. Davies was a lifelong oil man and a ranking executive of Standard Oil Company of California who was widely believed to be next in line for the presidency of the oil company. The duties of the

The United States was at war. When she heard the Division of Oil and Gas was looking for an inspector for the Taft office, Margaret Binkley, who held a degree in geology and paleontology, including courses in petroleum engineering, applied, realizing she was entering a man's world. (Margaret Binkley.)

new agency were to determine the increased amounts of petroleum products that would be needed for rearmament and to make specific recommendations for fulfilling these requirements.

A few days after Davies' appointment, Ickes and Davies at a meeting in Washington outlined their proposals for a "government-industry team" that would operate not on coercion but upon voluntary cooperation. The organization they described was to have a small Washington staff and five district operating divisions, managed by people of recognized ability and long experience in the oil business. They asked the oil industry to begin setting up the operating divisions at once. The industry quickly responded, setting up the first divisions in July. The active oil men appointed to the divisions agreed to serve without pay.

At the American Petroleum Institute's convention in San Francisco in November 1941, W. R. Boyd Jr., president, said, "The American petroleum industry has the knowledge, the ingenuity and the patriotic urge in the existing emergency, by working hand and hand with understanding and sympathetic government agencies, to meet every conceivable public and private need for petroleum products."

On the morning of December 8, 1941, members of the oil industry met with Secretary Ickes while Congress was declaring war on Japan. The Petroleum Administration for Defense now became the Petroleum Administration for War and, with industry representatives, plunged into the task of putting the oil industry on a wartime basis.

At a mass meeting of oil men in the Biltmore Hotel ballroom in Los Angeles on December 16, E. E. Pyles, an executive of Southwest Exploration Company who served as chairman of the Production Committee for District 5, covering the Pacific Coast states, Alaska and Hawaii, said production plans called for "efficient operations that would make petroleum and petroleum products available at the proper places to meet military and civilian needs and accomplish the most effective use of critical materials." In substance, the program called for the Production Committee to supervise the development and production of oil in California with due regard for efficient methods so that oil reserves and critical materials would be conserved.

In the wartime months that followed, the federal agency introduced the concept of maximum efficient rates (MER) for producing the state's flush pools, from which oil flowed without the aid of pumps. The MER, based on sound engineering practices, was defined as a rate of production for a pool which, if continued for six months, would not result in loss of ultimate production. By federal order, no pool could produce in excess of this rate.

With the introduction of the MER concept, it became necessary to classify all the pools in the state as to their geologic characteristics and reservoir dynamics. This task fell to the Conservation Committee of California Oil Producers, which traced its beginnings to 1929 when its predecessors had been formed to curtail the overproduction that threatened to drown the

E. E. "Ernie" Pyles, right, served as chairman of the Production Committee for District 5, covering the Pacific Coast states, Alaska and Hawaii. Reese Taylor, left, president of Union Oil Company, threw his company's resources into the effort to meet military and civilian needs. (William Rintoul.)

California oil industry in its own production.

Within days of the attack on Pearl Harbor, the sea war moved to California's coast. Richfield Oil Corporation's tanker *Agwiworld*, enroute from Los Angeles to San Francisco with a load of oil, was riding low in the water off Santa Cruz when a Japanese submarine surfaced and opened fire with its deck gun.

The tanker's skipper, who had survived a U-boat attack in the North Sea during the first World War, brought the unarmed tanker hard aport to face the submarine and ordered full ahead in an effort to ram the sub. Then, the tanker, narrowly missing its antagonist, tried to hide in a blanket of smoke from the funnel. The sub followed, firing its deck gun, but the sea was choppy and most of the shots went wild. With only slight damage, the tanker steamed safely into Santa Cruz harbor.

Richfield's *Larry Doheny* was not so fortunate. The tanker was steaming north when a Japanese submarine fired a torpedo that struck forward of the bridge, sending the tanker down in a sea of flaming oil with the loss of two members of the Navy gun crew and the ship's third mate and chief engineer.

Just before sunset on the evening of February 23, 1942, the war moved directly against California's onshore oil facilities. A Japanese submarine surfaced off the Elwood field west of Santa Barbara. As the sub cruised slowly along the coast, clearly visible in the twilight, the gun crew fired thirteen rounds of 5 1/2-inch shells against Elwood from a range of 2,500 yards, and then submerged. Damage from the bombardment was estimated at $500. The news of the attack upstaged President Roosevelt, who had just begun one of his fireside radio chats when the shelling started.

Though wells on piers jutting into the ocean presented an imposing appearance, production from the Elwood field, a 1928 discovery, had peaked in 1930 with an output of 40,000 barrels per day, and the field was in a state of decline when the United States entered the war. (Spence Air Photos.)

The attack was the first by a foreign power on the continental United States since the War of 1812. Except for a salvo against a coastal target in Oregon, it proved to be the only enemy attack against the continental United States during World War II.

One night after the Elwood attack, an Army balloon escaped its moorings and drifted over Los Angeles. Anti-aircraft crews opened up, pouring more than fourteen hundred rounds into the sky while nervous Angelenos rushed to their windows to watch, wondering if an invasion was about to begin.

The nation's war effort called for increased production from California's oil fields, even as the field operators dealt with manpower and material shortages.

The situation also called for flexibility and ingenuity. The 26-inch gas line that carried gas produced near Coalinga in the Kettleman North Dome field, as the Kettleman discovery was named, to San Francisco Bay users was converted into an oil pipeline to release tankers for war service elsewhere. The first oil went into the line at the beginning of June 1942. By the end of December, about 120,000 barrels per day of oil was being transported through the line.

To make up for the deficiency in gas supply to the Bay Area, the Rio Vista gas field was opened up. At Kettleman, gas production was curtailed to the extent possible, and the gas that was produced had to be injected into the Temblor zone. E. J. Kaplow, deputy supervisor for the Division's District 5, reported the injection rate was 52.3 million cubic feet per day, mainly in Eocene wells.

In the Division's District 1, the Federal Government in September 1942 took over a portion of the Playa del Rey field for use as an underground gas-storage reservoir. The purpose of the storage project was to ensure a supply of gas for defense plants and industry in Southern California. Under terms of the arrangement, Union Oil Company stored gas in cooperation with Southern California Gas Company. Before the end of 1943, almost two billion cubic feet of gas had been injected into the field for storage. In one 24-hour period in December of that year, Southern California Gas Company withdrew 36 million cubic feet from the reservoir, attesting to the success of the project.

The war effort received wide coverage in issues of *California Oil World*, both editorially and in pictures. Technical articles took on a new hue, expressed in such titles as "The Effects of Camouflage Painting on the Temperature of Plant Equipment," "Rehabilitation of Old Wells Aids War Effort" and "Standardizing Wire Rope and Lengthening its Life to Conserve Steel."

The publication gave prominent coverage to a May 1942 meeting in Los Angeles at which the U.S. Salvage Director called on the oil industry for salvage programs for "iron and steel scrap, nonferrous metals, cotton and woolen rags, paper, rubber and burlap."

A news story reported the presentation of an Army-Navy "E" award to Axelson Manufacturing Company, one of many such awards to be presented to various oil-tool manufacturers before the end of the war. Lieutenant Colonel A. R. Baird, who made the presentation at the Axelson plant, was quoted as saying, "This award is in recognition of the fact that you did a good job of producing machines, tools, gauges and other essential items for the armed forces of your

On a February evening in early 1942, President Roosevelt had just begun a nationwide fireside chat on the radio when a Japanese submarine surfaced near Bankline Oil Company's Elwood 89 lease and began shelling oilfield installations in the first attack by a foreign power against the continental United States since the War of 1812. (Long Beach Public Library.)

In the Trico gas field, John O'Conner, head gauger for Standard Oil Company, kept wells on production, supplying gas for military and civilian purposes. (William Rintoul.)

The Federal Government turned a portion of the Playa del Rey field into use as an underground gas storage reservoir early in the war to ensure a supply of gas for defense plants and industry in Southern California. (Spence Air Photos.)

country and her Allies...Remember that you are all soldiers on the production line of freedom."

In a subsequent issue, the magazine singled out Tide Water Associated Oil Company for "a unique demonstration of operational ingenuity" at Kern River. With wartime demand for oil soaring, the company was credited with putting the field back on production "after it had supposedly been depleted beyond economical revival." Tide Water had accomplished the feat "by the simple expedient of pumping water from the formation faster than it could accumulate, using powerful shaft-driven turbine pumps."

The photographs the magazine published reflected the war. One showed a battery of storage tanks guarded by soldiers, another a sandbagged wellhead, known as a Christmas tree, with the caption, "It will take a direct hit to wreck this Christmas tree." Another showed a tank farm with a forest of wooden derricks in the background and in the foreground, a tall steel tower. The caption read, "Oil company observation towers aid in spotting saboteurs." A photograph showing cattle grazing by steel derricks carried the caption, "Bovine critters invade California oil fields. While not as efficient as sheep in the role of animated lawnmowers and minimizers of fire hazards, cows do their part in the War Effort and are now encouraged to munch in derrick strewn pastures."

Another picture showed two attendants servicing a car at a service station. One was wiping the windshield, the other checking the pressure in the front tire. Both represented a departure. The caption read, "Attractive young women have made their appearance as service station attendants at a number of stations of Tide Water Associated Oil Company in Los Angeles and San Francisco—and there will be more."

Like others, *California Oil World* displayed a service flag. It appeared on the title page and under the heading, "In the Service of Our Country," and soon carried five stars, indicating five employees were in the service. Sprinkled through the pages of the magazine was the reminder, "Buy War Bonds."

When Margaret Binkley joined the Division of

With wartime demand for oil soaring, majors and independents returned idled wells to production in the Kern River field. Tide Water Associated Oil Company was commended for putting the field back on production "after it had supposedly been depleted beyond economical revival." (William Rintoul.)

Oil and Gas in 1943, her duties as an inspector involved watching to see that wells were drilled, completed and abandoned according to state regulations. She witnessed the cementing of surface casing and cement squeeze jobs to prevent contamination of oil, gas and water sands. She also witnessed water shut-off tests, the placement of abandonment plugs, and inspected blowout prevention equipment.

Since she worked for the state, she found that the stringent rules that governed how long women could work did not apply to her and that, like men, she could work much longer hours. Many weeks the 40-hour week was somewhat academic, and she found herself working closer to 80 hours.

What was it like being the only woman in what up to then had always been a man's world? In an interview with Susan Hodgson many years later, Margaret Binkley, then Mrs. Philip Pack (Bill Binkley died in an Alaskan air accident after he left the service), said, "I think what made it possible for me was that I was anxious to succeed and people were very helpful. And the men in the oil fields were just kind people. They were just good. To have had Gil Peirce as my first Division supervisor was additional good fortune.

"But, of course, it was a fortuitous time to break into the field. My husband was overseas. Everyone felt it was important to get on with the war. There was a manpower shortage, and it was clear I wasn't taking anybody's job.

"Few really young men worked in the oil fields then. They were in the armed services or headed for greener pastures in shipyards or in other wartime projects. The oilfield men I dealt with were a special kind of people with a special kind of pride. Some had little formal education. But you don't work your way up from rig hand, to driller, to toolpusher, to drilling foreman or drilling superintendent or to other positions of trust by being behind the door when God passes out brains.

"Oilmen preferred to help people who make an effort to help themselves. They worked hard and expected other people to do the same. They had a strong set of moral values and their own rules of acceptable behavior. But, they were kind. They could have run me off, and they knew it. But they

A forest of wood and steel derricks dotted the treeless Buena Vista Hills near Taft, where Margaret Binkley during World War II became the Division's first woman engineer. (Division of Oil and Gas.)

didn't.

"I don't think being a woman was an advantage. Once the men were sure it wouldn't hurt my feelings, they'd say things like, 'We don't approve of women working, and we don't approve of women working in the oil fields, but you're different.' But, of course, I wasn't different. They just knew me; that was all. Eventually, I think we just took each other for granted. I thoroughly enjoyed those days. They were a marvelous experience."

In her work, Margaret met few other women professionally. "During the war," she said, "some oil companies hired women as pumpers, gaugers and for other jobs traditionally held by men. In addition, I knew of one other female geologist nearby, a woman from Berkeley who worked for a major oil company, but I never ran into her in the field."

The same year she joined the Division's Taft office, notices to drill new wells in District 4 reached an all-time high with 849 new notices, compared with 416 in 1942. Most of the activity took place in the Midway-Sunset and South Belridge fields, which fell within the jurisdiction of the Taft office. Abandonments, too, almost doubled the preceding year's, representing an effort to salvage casing for use in drilling more wells.

"I don't think I've ever worked as hard. And we were so busy. We had about 40 rigs drilling, 17 in Elk Hills alone. The West Side of the district was a big area. It'd be nothing unusual to make two or three trips to Elk Hills, and probably go to Belridge or Lost Hills and all around in one day. "Of course, there was nobody on the road except you and a few others. Gil Peirce ran the whole West Side for awhile with one efficient secretary, himself and me. Sometimes, we had another engineer, either ours or borrowed from another office. There was all this work to be done, and you really felt needed. Whether you were or not might be debatable, but that's what we believed."

To get the field experience she acquired during the war, she reflected, would have taken perhaps 10 years under other circumstances. "We had so many companies drilling, using every type of rig including cable tools. I was on all of them, many times. We just chased so many holes. It isn't that I was all that smart. But everyone was very helpful, and I did work hard. It was fun."

While most of the wells that were drilled in the Taft area she covered were development wells, at least one was an exploratory well that commanded some attention.

The well was Standard Oil Company's KCL No. 20-13 near the South Coles Levee field. In the fall of 1944, the well returned the world's depth record to California from Texas, where a Phillips well had bested the Continental well at Wasco. The Standard well passed the Phillips depth of 15,279 feet and kept on drilling ahead. As drilling continued, *California Oil World* said of the world's first three-mile well, "Little if any of the equipment is new, and much of it is not of the most recent design," reflecting wartime circumstances under which the oil industry operated. Standard continued drilling to a depth of 16,246 feet, but failed to develop production.

The U. S. Navy, as administrator for Naval Petroleum Reserve No. 1, opened up the Elk Hills field, putting shallow zone wells on production to help the war effort. (William Rintoul.)

Under the impact of the war, the Navy initiated an active drilling program in Naval Petroleum Reserve No. 1, which blanketed the Elk Hills field. The result was discovery of production in the Stevens sand, which proved to be the reserve's largest reservoir. Above, a Stevens zone well. (William Rintoul.)

When the war ended in 1945, California's production for the year totaled 328.3 million barrels, representing a gain of 98.6 million barrels over the 229.7 million barrels produced in 1941 when the nation went to war. Through the war years, the state had produced one billion barrels, making a solid contribution to the approximately six billion barrels the United States supplied the Allied forces during the war. The American share represented about 85 percent of the total of almost seven billion barrels used by the Allies.

In a speech before the American Petroleum Institute in Chicago in November 1945, Ralph K.Davies, Deputy Administrator, Petroleum Administration for War, read a letter from the Army-Navy Petroleum Board that said, "At no time did the Services lack for oil in the proper quantities, in the proper kinds, and at the proper places...Not a single operation was delayed or impeded because of lack of petroleum products. No Government agency and no branch of American industry achieved a prouder war record."

In California, Margaret Binkley continued her career as a petroleum geologist in the Division's Taft and Bakersfield offices until 1953, when she joined Oceanic Oil Company, an independent oil company in Bakersfield.

Wartime production from the Huntington Beach field climbed to its highest level since the boom days of the 1920s, reaching 47,000 barrels per day before the war's end, an increase of 21,000 barrels daily from prewar production. (Los Angeles Public Library.)

7 From Confusion Hill to the Subsidence Bowl

Wildcatters had begun looking for oil in Cuyama Valley almost from the earliest boom times in the neighboring San Joaquin Valley. Faintly outlined against Chalk Mountain, photo left, is the rig with which Norris Oil Company on New Year's Day 1948 proved up Cuyama's first commercial field, quickly drawing Richfield Oil Corporation, represented by the rig on the right, into the play. (William Roth.)

No one knew what would happen when the war was over. There were dark prophecies of a return to depression. Many economists believed that the frantic industrial production of the war years would falter and fade with the dismantling of the war machine. But when governmental controls were lifted, the market that had been restrained for five years suddenly boomed.

The American people set records for births, building and buying in the postwar years, creating new demands for the products of all industries, particularly petroleum. Any decrease in demand that came when tanks were scrapped and bombers mothballed was more than offset by the growing number of cars that poured out of Detroit. In California's oil fields, there was only a short drop off during the adjustment period from war to peace. The state's wartime production had peaked in 1945 at 328.3 million barrels, averaging approximately 900,000 barrels per day. In 1946, production dropped to 317.1 million barrels, or an average of about 869,000 barrels daily. The following year, California's producers topped the wartime peak with an output of 336.7 million barrels, averaging 922,000 barrels per day.

During the war, the top year for drilling activity had been 1944, when California operators filed notices for 2,252 new wells. In the first year following the war, the Division of Oil and Gas processed 1,809 new well notices, a decline of about 20 percent. The turnaround came the following year with 2,127 new well notices. In 1948, operators filed notices for 2,772 new wells, representing an increase of 23 percent over the wartime high. The transition from World War II into a booming peacetime economy seemed complete.

Wildcatters fanned out over California, proving up giant fields in two new provinces and finding more oil in the aging, nondescript Placerita field only seven miles from where the state's first commercial well had been completed in 1876. The Placerita discovery would go on to cause the Division of Oil and Gas more than its share of concern in the 1950s.

One of the unproved provinces that caught the attention of postwar explorationists was the fertile Salinas Valley, hemmed in on one side by the Gabilan Range and on the other by the rugged Santa Lucia Range. The valley was made famous in literary circles as the birthplace of the controversial author John Steinbeck and the setting for some of his best known works, including the novel *Of Mice and Men* and the short story *The Red Pony*.

The Salinas Valley had been a target for wildcatting interest at least as early as 1908 when Standard Oil Company had leased 20,000 acres on both sides of the railroad and the Salinas River between San Ardo and Bradley. The company then announced plans to drill an exploratory hole with a rotary rig newly arrived in California from Louisiana with a nucleus of crewmen experienced in manning the rotary, which then was largely unproved in California. The company had moved the rig to the drill site, but before the well could be spudded, management decided against drilling the hole and moved the rig farther north to the Altamont Pass between Oakland and Tracy to drill what proved to be a dry hole.

Standard eventually did open an exploratory campaign in the valley in 1929, but the hole was a duster at 6,093 feet and the company moved on to other prospects. Before and after the drilling of the unsuccessful wildcat, other wildcatters had tried their luck in the valley, drilling and abandoning at least 40 exploratory wells before the post-World War II years.

Among those who tried and failed, some with more than one well, were Continental Oil Company, Shell Oil Company, Union Oil Company, Tide Water Associated Oil Company and The Texas Company, later Texaco Inc.

Texaco initiated its campaign in 1936 seven miles northwest of King City with a wildcat, but the well proved dry. Before World War II interrupted the search, the company drilled and abandoned six more wildcats, extending the hunt to unproved ground southeast of King City.

After the war ended, Texaco resumed its exploratory effort with a wildcat on the Aurignac property, four miles southeast of the community of San Ardo. The first well encountered heavy oil just below 3,000 feet, but the sand was wet. The company spudded in to drill a follow-up well. The surface casing parted. Texaco abandoned the hole in favor of drilling a twin well, which went to 3,165 feet, finding heavy, tarry oil. The company was not able to complete the well as a commercial producer because of the large volume of sand that entered the hole.

Texaco moved two miles to the northeast to try again. The Lombardi No. 1 cored saturated oil sand just below 2,100 feet and was completed in November 1947 as the discovery well for the San Ardo field, pumping 155 barrels per day of clean 10-gravity oil. Texaco produced the well for two days and shut it in. A follow-up well drilled by Jergins Oil Company three miles to the southeast was completed as a gas well.

The discovery at San Ardo did not touch off immediate excitement. Of the find, *California Oil World* reported, "Texaco's success marked the end of an extensive oil hunt in the great Salinas Valley...While the discovery well and the follow-up job failed to find oil in quantity, they did serve in proving the existence of petroleum in hitherto unproductive Monterey County."

It would not be until the 1950s that the field would begin to come into its own as an important supplier of California crude with recoverable reserves estimated in excess of 500 million barrels of oil.

Another unproved province that drew interest in postwar years was a high, arid valley in the erratic, multihued mountains of the Sierra Madre Range between the San Joaquin Valley and the coastal city of Santa Barbara. Nomadic American Indians had named it Cuyama, their word for *clam*, after the fossil clam shells they found there on treks through the pass from the San Joaquin to the coast. Once a Spanish land grant, Cuyama had been settled by a handful of persevering homesteaders, who made a spare living raising cattle, alfalfa, sugar beets and seed potatoes.

The valley had been a target for wildcatters almost from the earliest oil booms in the neighboring San Joaquin Valley. The signs that brought the first oil prospectors to the San Joaquin Valley also existed in Cuyama Valley. There were oil seeps and provocative outcrops. There the similarities seemed to end. Various companies and individuals had drilled Cuyama without success, giving the valley the reputation of a wildcatters' graveyard. One major company went so far as to pronounce Cuyama Valley a place whose

The Salinas Valley had been a target for wildcatters since at least as early as 1908. After World War II ended, Texaco resumed the search that had been ended by the war and capped the campaign with discovery of the San Ardo field, a giant field. (William Rintoul.)

geology would seem to preclude the existence of any major oil accumulations.

Two searchers were not quite ready to give up. They represented the dominant elements of oil exploration, one a major oil company, the other an independent wildcatter. The approaches were tailored to the character of the two parties. Though dissimilar in style and resources, they both featured strongminded individuals willing to gamble on their faith.

The major oil company was Richfield Oil Corporation, which had survived receivership in the dark days of the Depression and come back under the leadership of a new president, Charles S. Jones, to make a major discovery in the North Coles Levee field in the San Joaquin Valley. The company was willing to flex its exploratory muscles in places others were not so willing to tackle. An on-the-ground survey of Cuyama Valley convinced the company's exploration team that Cuyama was a sleeper.

The independent was an agile, silvery-haired oil prospector named George Hadley who had spent nearly 50 of his 71 years looking for oil from Canada to Mexico. Hadley had come to Cuyama just before the war in a beat-up car that served as his home while he scouted the valley from one end to another, becoming ever more convinced that it was prime oil-hunting ground.

Looking for financial backing, Hadley found it in the form of Halvern L. Norris, a scholarly, middle-aged bachelor who lived in Ventura and interested himself in Oriental letters. Norris, a retired vice-consul to Siam, later Thailand, and a veteran of many years' service in United States legations in Japan, China and Yugoslavia, among other countries, agreed to put up money for a well. Hadley and Norris enlisted the support of Arthur Scott, a Maricopa driller in his late thirties who had spent the greater part of his working life on the floors of drilling rigs.

The first two wells the Hadley-Norris-Scott group drilled were dry. The campaign was a poor-boy operation, with times when the crew on the rig consisted only of Hadley and Scott instead of the five men who normally would have been there. The period was a difficult one for Norris Oil Company, which survived three reorganizations, including a time when some of the company's stock changed hands for only eight cents a share, with proceeds going to pay hotel rent.

In November 1947, Norris Oil Company began drilling its third exploratory well close to Chalk Mountain. On New Year's Day 1948, Norris completed the well that flowed 190 barrels per day of 21-gravity oil, proving up the Russell Ranch field as Cuyama Valley's first discovery.

The find forced Richfield's hand. The company had not intended to drill until later in 1948 and had kept its plans secret. When news of the Norris discovery reached the company in Los Angeles, Richfield mounted an all-out effort to secure leases. By the middle of January 1948, the company had acquired over 150,000 acres, representing in light of future events almost 90 percent of the potential production in Cuyama Valley.

Three men who played major roles in the discovery and development of Cuyama Valley's oil were, left to right, Charlie Jones, president of Richfield Oil Corporation; Frank Morgan, who directed the company's exploratory effort; and Rollin Eckis, a key member of the exploration team. (William Rintoul.)

While Norris battled water problems at its discovery well, Richfield moved in to drill its first wildcat three miles to the southeast. Seven weeks after spudding in, the company completed Russell A No. 28-5, which flowed 508 barrels per day of 38-gravity oil. Richfield quickly scored again with a wildcat two miles to the northwest, getting a flow at a rate of 3,041 barrels per day. There could be no doubt that Cuyama was an oil province.

Richfield put eight rigs to work. Within a month, the company was completing a well every four days. Service company representatives proliferated to meet the demands of the drilling boom. One was a 25-year-old recent graduate of Yale who sold Security bits as part of his training program with Dresser Industries, the parent company of Security Engineering. His name was George Bush, and he would become President of the United States.

Production from the Russell Ranch field rapidly approached 20,000 barrels per day. To solve the problem of getting the oil to market, Richfield bought thirty miles of battered and dented secondhand six-inch steel pipe that had been used by the Army as a surface pipeline string across Europe to supply fuel for the tanks of General George S. Patton's Third Army. With the makeshift line, Richfield was able to start moving oil to market almost immediately by means of a tie-in with existing San Joaquin Valley pipeline connections.

From Russell Ranch, Richfield moved five miles to the southeast to drill a wildcat on a parcel that Glenn Homan of Los Angeles had leased from the Federal Government for $170. The company completed Homan A No. 81-35 in May 1949, flowing at a rate of 5,000 barrels per day from a total depth of 4,392 feet. The well proved up the South Cuyama field, a giant that subsequently would produce more than 200 million barrels of oil.

While the Cuyama boom held center stage, the completion in April 1948 of a shallow well by Nelson-Phillips Oil Company in Placerita Canyon some twenty miles from downtown Los Angeles occasioned passing interest because of the 70 barrel-per-day well. Though hardly a barn burner, it offered hope of being the first good well in a field that had been discovered 28 years before and since then had produced only intermittently. Interestingly, the Placerita field was only seven miles from the Pico Canyon well that California Star Oil Works Company had completed in 1876 as the first commercial oil well in California. This well, Pico No. 4, was still producing when Nelson-Phillips brought in its well, and continued producing until 1990, when it was plugged and abandoned by Chevron USA Inc. after 113 years of

Bringing in a big well in the South Cuyama field was a cause for celebration by Richfield Oil Corporation. (William Rintoul.)

production, a state record.

Completion of the Nelson-Phillips well in 1948 started a drilling campaign that by the end of the year resulted in the completion of 23 wells producing 450 barrels per day at depths ranging from 500 to 1,500 feet.

The Placerita activity loomed large to M. R. Yant, who had drilled the area without much success in the 1930s and subsequently had subdivided and sold portions of his property in small lots, some no larger than one-tenth of an acre. Yant, working as an electrical contractor in Hollister when he learned of the new find, interested Ramon Somavia, a rancher, into backing an exploratory well on the brush-covered hill that he had subdivided. The subdivision that Yant had envisioned, like so many others of its time, existed on paper only. In early March 1949, the Somavia well, designated Juanita No. 1, was completed flowing 340 barrels per day of 22.8-gravity oil from an interval at 1,737 to 1,830 feet. The well was the best yet for the Placerita field, and it proved up an entirely new producing area with higher gravity oil than any found before.

The discovery touched off a scramble for leases. The first finding on the part of speculators looking for acreage, not surprisingly, was that the hill where Somavia had found oil had been highly subdivided. Some who

Confusion Hill, when the rigs were there. (Ira Carroll, Petroleum World.*)*

When the rigs left, the hill that M. R. Yant had subdivided looked like a tank farm. (Pacific Oil World.)

At Confusion Hill, Atlantic Oil Company completed Lockwin No. 1, marked by the Christmas tree, left center. A. G. McHale moved in to drill an offset well, Woodworth No. 1, right, on a lease measuring 33 by 61 feet. Gordon Drilling Company offset with Peggy Moore No. 10, rig on left. (Ira Carroll, Petroleum World.*)*

purchased lots from Yant had, in turn, subdivided their property, reducing parcels to as little as one-twentieth of an acre. The next discovery was that Yant's property had not contained 80 acres. The section lines were found to be off, as the 80 acres that Yant had subdivided actually were only 71 acres. Consequently, all the parcels bordering on the exterior lines of the supposed 80-acre property were shortened by the overlap in deeds. It also developed that in many cases the legal titles were tangled, with as many as three claimants for some of the tiny parcels.

Obviously, the oil that lay beneath the subdivided area recognized no surface boundaries. It simply lay in the earth, waiting to flow to the surface in greatest volume to those who tapped it first. Operators rushed to drill, mindful that each day's delay meant a loss of production to neighbors quicker on the draw.

The hill became "Confusion Hill." More than two dozen drilling rigs crowded into the 71-acre space, leaving drilling crews, service and supply men, speculators and landowners with visions of overnight wealth to wander confusedly through the maze. In the charged atmosphere, some wells came in for initial productions as high as 3,000 barrels per day.

A concerned observer of the frantic congestion was R. D. Bush, State Oil and Gas Supervisor for the Division of Oil and Gas. Bush sought to enforce the provisions of the Spacing Act embodied in the *Public Resources Code of the State of California*. The act provided that no more than one well be drilled per acre. With operators rushing to bring in wells on whatever size lot they could acquire, the constitutionality of the act was soon challenged. In the case of *People* v. *Metcalfe Oil Company*, the state attempted to get an injunction to prevent the drilling of a well. The operator opposed the action. In September 1949, Superior Court Judge Clarence M. Hanson denied the application for an injunction, pending trial. The decision opened the way for an even more intense drilling campaign. The number of rigs working on Confusion Hill climbed to forty-eight.

In October, production peaked at 36,000 barrels per day from more than 100 wells.

One of the Clyde Hall Drilling Company crews that worked the boom in the Placerita field included, left to right, B. J. Spears, lead tong; Wilson Morris, cathead; E. A. Pitts, derrick; and Ray Gossett, driller. (Pacific Oil World.)

Accompanying gas was blown to the air. The Division warned operators that unless steps were taken to eliminate the waste, the Attorney General would be requested to take action. Operators met, and for a time it appeared a voluntary curtailment plan would be successful. When it became apparent no results were being obtained, Bush requested the Attorney General to prepare a case charging 21 corporations and 19 partnerships with "wastefully causing and permitting natural gas and natural gasoline contained therein to blow, release and escape into the air."

In response to the Division's action, Judge Hanson issued an order requiring operators to show cause why a temporary restraining order should not be granted. Negotiations between interested parties seeking an outlet for the field's gas met with failure. The gas had a high carbon dioxide content and a low heating value. A study indicated a plant for the recovery of the gasoline alone was not feasible because the gas had a low gasoline content. The gas wastage injunction action was placed off calendar in February 1950, and no further steps were taken to stop waste. It was estimated that operators had blown 2.6 billion cubic feet of gas into the air during the last six months of 1949. By the end of 1950, most of the drilling rigs were gone and production had dropped to 12,230 barrels per day, which was barely one-third of what it had been a year before. Therefore, no further legal action was taken.

In the San Joaquin Valley, operators looked for deep oil. On August 20, 1953, The Ohio Oil Company broke the world's depth-record of 20,521 feet and continued to drill ahead at KCL-A No. 72-4 in the Paloma field. The well reached a final depth of 21,482 feet in October of that year and was abandoned at the end of 1954 after exhaustive testing failed to produce commercial production. The $2,250,000 Paloma deep test proved to be the last world's depth-record well for California. In 1955, a Richardson and Bass well in Plaquemines Parish, Louisiana, broke the Paloma record and continued on to 22,570 feet. The record would be broken many times in subsequent years, finally by a pair of wells in the Anadarko Basin near Elk City, Oklahoma, that

Some 25 years after the Division of Oil and Gas had its beginning, a group posed for another family portrait. The 1940s group included, left to right, S. G. Dolman, deputy supervisor, District 3; R. D. Bush, supervisor; H. V. Dodd, deputy supervisor, District 4; E. H. Musser, deputy supervisor, District 1; C. C. Thoms, deputy supervisor, District 2; Ed Kaplow, deputy supervisor, District 5; E. Huguenin, chief deputy supervisor; and R. W. Walling, later at various times, deputy supervisor, Districts 1, 4 and 5. (Division of Oil and Gas.)

went below 30,000 feet. Lone Star Producing Company's Baden No. 1-28 in Beckham County bottomed at 30,050 feet in 1972. The same operator's Bertha Rogers No. 1 in Washita County went to 31,441 feet in 1974. Neither developed commercial production at depth.

While some operators were willfully blowing gas to the air at Placerita and others were looking for deeper production in the San Joaquin Valley, many were busy in the Sacramento Valley looking for more gas fields and trying to enlarge on the discoveries that had been made starting in the 1930s. In the first few years after the war, there were a number of new gas field discoveries, among them Denverton, Dunnigan Hills, Durham, Freeport, Pleasant Creek, River Island, Wild Goose and Winters fields. And there were extensions and new pool finds in existing fields. The drilling pace picked up until, in 1950, the Rio Vista gas field was the most active field in District 5, accounting for 31 new-well notices.

As a consequence of increased activity in the Sacramento Valley, District 5 was divided into two districts in 1955. The original District 5 retained

Driller W. T. White with diamond bit that set a new world's depth record at 20,522 feet in the Paloma deep test. This is how the bit looked when it was pulled from the hole at 20,531 feet. (Marathon Oil Company.)

On a hot August day in 1953, The Ohio Oil Company's KCL-A No. 72-4 became the deepest well in the world, the last in California to hold that distinction. In the oil patch, the exploratory well was widely known as the Paloma deep test. (William Rintoul.)

Fresno, Madera, Kings, Mono, Mariposa, Merced, Stanislaus and Tuolumne Counties. The new District 6 covered Sacramento, San Joaquin, Solano and all other counties not included in other districts. Headquarters for both districts remained in Coalinga, with Eugene Murray-Aaron serving as district deputy for both. In 1957, a separate office for District 6 was established in Woodland, and Fred Kasline was named district deputy.

Another change occurred in 1955 when the Conservation Committee of California Oil Producers, which had continued setting maximum efficient rates for California's flush pools after the war ended, was given quasi-regulatory status by the State Legislature as the designated advisor to the State Oil and Gas Supervisor on all matters involving conservation of the state's oil and gas resources. The enabling legislation also gave state sanction to the recommendations of the Conservation Committee, requiring that the Committee's recommendations be delivered to the State Oil and Gas Supervisor and be made open to public inspection. Compliance with the recommendations continued to be voluntary.

The "Case of the Disappearing Land" posed a serious problem for Navy facilities in the Wilmington field. (Long Beach Public Library.)

Meanwhile, though the war had ended in 1945, mysterious events in the Long Beach Harbor area made it appear that an unseen foe had drawn an invisible target over the area and placed the Navy's $75 million shipyard almost precisely in the center.

The walls and foundations of buildings began to crack. Sewer lines broke. Storm drains ruptured. The railroad tracks that served the yard began to buckle. When surveyors made the annual August survey of Long Beach Harbor Department bench marks, they found that elevations were less each year. In one year alone—1951—one bench mark near the Navy's shipyard dropped more than two feet. The inescapable conclusion was that the land in the harbor area slowly was sinking beneath the sea.

As land sank, the Navy like others with facilities in the area took what remedial actions were necessary to protect installations, trucking in landfill, raising foundations and constructing dikes. Subsidence continued until the Navy's shipyard lay below sea level, protected from flooding only by dikes, some as high as twenty feet. There was talk of abandoning the yard and moving the Navy operations to some undisclosed site somewhere else on the Pacific Coast, taking with it about one-seventh of the Long Beach area's economy.

The Navy was not the only one affected by disappearing land. Going down with the Navy in the subsidence bowl was a conglomeration of factories, shipyards, docks and warehouses that employed more than 1,000 people representing an annual payroll of more than $6 million. The City of Long Beach also

Subsidence caused the northern portion of Pier A in the Long Beach Harbor to sink so far that at times ocean water washed over the wharf. The picture shows Berths 1, 2 and 3 in February 1958 when the surface level had sunk about ten feet. (Long Beach Public Library.)

was threatened. Streets buckled, sewers and underground pipes broke, the towers of the Commander Heim Bridge tilted and the city hall moved three feet closer to the center of the bowl.

As if all of that were not enough, the future of California's largest oil field also was at stake. The outline of the area that was sinking coincided with the heart of the Wilmington field. The field not only was California's number one field but also the second largest in the United States, runner-up only to the East Texas field.

Oil operators battled to preserve production. The sinking land's surface was only part of the problem. Beneath the surface, shearing action caused by the subsidence crushed casing and, in some cases, sheared it off, causing loss of production. A survey of some 160 wells found deformed casing in 96 and damage so severe in 22 that operators had no choice but to abandon the wells. The problem threatened not only the developed portion of the field but also untapped tidelands extending eastward, which held hundreds of millions of barrels of oil that might never be tapped.

Among those making studies at Wilmington was Richfield Oil Corporation, which operated wells in the field and had valuable marine terminal facilities near the center of the subsidence bowl. The company's study linked oil production with the cause of sinking land, citing three characteristics of Wilmington's pools as the culprits.

First, unlike other oil reservoirs in California, which were strongly folded with dips of up to 45 degrees on the flanks making strong arches, the Wilmington field's reservoirs were relatively flat, with dips of only 10 to 15 degrees on the flanks.

Second, Wilmington's reservoirs were relatively close to the surface. Instead of being overlain by thick layers of shale that occurred in most oil fields, they were overlain by thin layers of shale and by gravel and sand. And finally, the Wilmington reservoirs were extremely thick, but at the same time they were not well cemented or consolidated. When reservoir pressure was reduced, there was a greater than usual volume of material subject to compaction.

In August 1955, Charles S. Jones, Richfield's president, stood before the Rotary Club of Long Beach at a time when the center of the subsidence bowl was more than 21 feet below the former surface of the ground and told listeners if nothing was done, the city, its industrial base in the more than 10-square-mile affected area and even such new developments as Belmont Shore, Naples and Alamitos Bay faced a grim future.

Jones proposed a solution. His proposal, instead

Subsiding land seemingly put the ship in the background on a collision course with a Long Beach Harbor Department office protected by earthen dikes. (Atlantic Richfield Company.)

of stopping production or spending more money to build dikes and take other remedial measures, would result in far greater revenue to the city and oil producers over a longer period through the recovery of far more oil than they would recover otherwise.

The solution was to begin a waterflood. By injecting water under pressure behind the oil-water contact, great pressure would be added to the water and it would act like a giant piston pushing oil ahead of it to the areas of low pressure where the oil could escape through wells. In addition, the injected water would increase the reservoir's pressure, thereby arresting the subsidence.

For waterflooding to succeed, each oil- and gas-reservoir in the field would have to be operated as a unit. In the Wilmington field, there were thousands of different surface owners and more than 100 different operating interests. Under existing law, it would be impossible to have a unit operation unless there was an agreement to which 100 percent of all of the owners were parties.

It was obvious that legislative help would be required. The City of Long Beach sought state assistance in enacting special legislation. The State Assembly's Interim Committee on Manufacturing, Oil and Mining Industry held two public hearings late in 1957, including one in Long Beach and another in Los Angeles. From the hearings and study of the testimony, the Committee concluded the only feasible method to arrest or reduce subsidence was by repressuring the oil zones, and that compulsory unitization might be necessary to accomplish that goal.

On March 4, 1958, a charter election was called in Long Beach to seek an amendment to the charter permitting the city to enter into unitization agreements. Voters favored the amendment 15 to 1.

Later in the same month, Governor Goodwin J. Knight called a special session of the Legislature to consider the subsidence problem. The Legislature enacted the Subsidence Control Act of 1958, effective July 24. The act encouraged voluntary pooling and unitization, but provided for compulsory unitization if necessary. It provided rules for determining costs of initiating and conducting a repressuring operation and set up procedures for filing plans of operation. It also provided for hearings and for

This stainless-steel pump was lowered into a water-source well for delivery of salt water to be used in flooding designed to halt subsidence and increase oil recovery at Wilmington. (Port of Long Beach.)

appeal or judicial review of cases where interested owners objected to the formation of a unit.

In the event that voluntary programs might fail, the State Oil and Gas Supervisor was empowered to order unitization if the order was acceptable to working-interest owners entitled to 75 percent of the proceeds of production from the proposed unit. If ratification of the order could be achieved from persons entitled to 75 percent of the proceeds of production in the proposed unit area, a city or county might exercise the right of eminent domain to acquire properties of nonconsenting persons.

The Federal Government made it plain it would not tolerate any delaying action. On August 15, 1958, the government filed in the U.S. District Court a $54 million property damage suit and an injunction to compel the City of Long Beach, the State of California and oil operators in the Wilmington field to take immediate steps to halt subsidence in the Long Beach Naval Shipyard and other governmental installations in the area.

The stage was set for all-out war on subsidence. To determine the boundaries of the area to which the Subsidence Control Act would be applied, State Oil and Gas Supervisor E. H. Musser held public hearings in September in Los Angeles. Following the hearings, Musser established exterior boundaries of lands comprising the subsidence area at or near the limits of the Wilmington field, including the undeveloped eastern extension into the tidelands. The subsidence area comprised approximately 21,600 acres, including all of Long Beach Harbor and the eastern portion of Los Angeles Harbor, extending to the southeast as far as the Orange County line, to the north to Lomita Boulevard and to the west as far as Figueroa Street.

To accelerate the control of subsidence in critical areas, six fault block unitization areas were formed rather than one fieldwide unit. The top priority was given to the Long Beach Naval Shipyard and the area underlying downtown Long Beach. Producers in the vicinity of the shipyard did not wait for final unitization agreements before beginning water injection.

The Wilmington field quickly became the busiest field in California. Eight rigs went to work. Instead of oil wells, they drilled water-injection wells for what soon would become the world's largest waterflood project.

The City of Long Beach undertook the responsibility of constructing the necessary water plants and supply systems, spending some $6 million in the work to further encourage private parties to participate by lowering their initial capital outlay. The city would recover its investment through the sale of injection water.

On October 29, 1959, the Long Beach Harbor Department hosted a half-day event to celebrate the greatest stride that had yet been made in development of the huge waterflood—the dedication of the Terminal Island and Mainland water-injection plants.

At a given moment, Captain C. J. Palmer, Commander of the Long Beach Naval Shipyard, turned a valve and water arched into the air from a standpipe 20 feet to the rear of where the Captain stood.

This was "Big Squirt," a dramatic portrayal of the stream of water that would be pumped into the ground to stop the sinking of the City of Long Beach, its people and its multimillion-dollar industrial complex.

The water gleamed in the afternoon air. As it

mounted hundreds of feet into the sky, it formed a spray, and the fragment of a rainbow appeared, nature's good omen.

Five months later, U.S. Attorney Laughlin Waters announced that in view of the satisfactory progress being made in the $30 million Wilmington waterflood's control of subsidence in the shipyard area, the government's injunction suit against oil operators would be kept off the court calendar.

In 1964, the Wilmington field produced its one billionth barrel of oil, becoming the first oil field in California to reach that mark and the second in the nation, following the lead of the East Texas field.

Even as production was reaching that milestone, operators were injecting half a million barrels of water a day into the field, or five times more water than the field's average production of 97,000 barrels per day of oil.

With subsidence under control, the way was clear to unlock a treasure some had feared might be lost forever. The City of Long Beach and the State of California, as co-owners of mineral interest, in February 1965 opened the seaward extension of the Wilmington field to competitive bidding.

They scheduled six days of sales for varying interests in a 4,600-acre tract in the East Wilmington area believed on the basis of geological studies and core drilling to contain one billion barrels of recoverable oil.

The first day's sale offered the choice prize, an 80 percent interest in the tract. More than 200 persons jammed the Long Beach City Council chambers and others watched proceedings over closed-circuit TV in the Veterans Memorial Building auditorium one block away while two competing bids were opened. The winning bid came from a five-company group that offered 95.56 percent of net profits for the right to develop the tidelands extension of California's largest oil field.

The successful bidding group included Texaco Inc., Humble Oil & Refining Company, Union Oil Company of California, Mobil Oil Corporation and Shell Oil Company, known by the acronym formed from the first initial of each company's name as the THUMS group.

The remaining 20 percent working-interest shares of the tract were split into five offerings sold on succeeding days to Pauley Petroleum Inc.-Allied Chemical Corporation and Standard Oil Company of California-Richfield Oil Corporation for offers ranging from 98.277 percent to 100 percent of net profits.

Why had oil companies bid so high? One reason was proximity of the oil reserves to refining and transportation facilities in the center of the largest petroleum-consuming region in the United States. Another was that the undertaking involved little financial risk. From profits, the field contractor would be reimbursed for investment and operating cost and receive a 3 percent management fee.

For THUMS, submission of the winning bid meant perhaps the brightest prospect ever to face a newcomer to California's operational ranks. From a company without production at the beginning of 1965, the consortium was able to announce confidently that within three years it would be among the top ten oil producers in the United States.

The THUMS group quickly formed THUMS Long Beach Company with a cadre of employees on loan from parent companies, supplemented by a number of individuals employed directly. On July 16,

1965, some six months after the sale, THUMS spudded in to drill its first well. The well, sited on Pier J, was to be the first of 1,100 wells scheduled to tap the tidelands.

Before the year ended, THUMS began construction of two of four islands off the City of Long Beach in water depths of 30 to 40 feet from which the major share of the tract would be developed. A huge dredge, driven by a 10,000-horsepower electric motor, sucked up sand from the bottom of the harbor to form bases for the islands. Barges hauled rocks for the perimeters from a quarry on Catalina Island some 22 miles away. The dredge filled the centers.

Drilling began from Island Alpha in 1966 and from Islands Bravo, Charlie and Delta in 1967, on schedule. By August of that year, 19 rigs were drilling from the four islands and Pier J, making East Wilmington the most active oil development in the United States.

In an effort to ensure that the offshore operation did not offend sensibilities, THUMS constructed drilling towers with pastel balconies to look like office buildings or penthouse apartments. Each tower had a triple deck substructure to house mud pumps, shakers for drill cuttings, power converters, motors and other equipment. Mounted on rails, the towers traveled around the islands, and slant wells were drilled only six feet apart at the surface.

The islands were planted with semitropical Mexican fan palms and Canary Island date palms. For lower plantings, climate-resistant sandalwood trees, saltbushes, oleanders, acacias and Moreton Bay fig trees helped beautify the operation. An elaborate water and fertilizing system was installed.

To top it all off, three spectacular 30-foot-high waterfalls were added. The falls were visible from

Rigs camouflaged to look like office buildings and extensive landscaping with palms and other semitropical trees on the THUMS islands made the seaward extension of the Wilmington field an attractive neighbor. (Department of Oil Properties, Long Beach.)

around the bay, offering a vista so pleasing that city fathers demanded, and got, floodlights for night viewing.

By the end of 1967, THUMS was completing wells at a rate of 350 per year, or almost one a day. Production had climbed past 50,000 barrels per day and was continuing to rise.

That same year, the Long Beach City Council approved a new name for the group of islands, calling them the Astronaut Islands. Individually, they were to be known as Grissom, White, Chaffee and Freeman, in memory of the astronauts who lost their lives in the U. S. space program.

Even as the THUMS development program pushed the Wilmington field's production toward a peak of 224,135 barrels per day in 1970, the Division in its annual report for 1968 said the entire Wilmington-Long Beach area, still undergoing massive water injection, seemed to be experiencing a regional uplift, with the maximum rebound being four inches. The bench mark on Terminal Island that marked the point of maximum subsidence of approximately 29 feet gained one inch in elevation during the year. It marked the first time that the cumulative subsidence figure had reversed its trend.

8 Down to the Sea

Summerland oil field, circa 1904, near Santa Barbara. The first pier was built in 1897 to serve as a base from which to drill the nation's first offshore well. (Seaver Center.)

The land that was subsiding at an alarming rate in the 1950s in the Wilmington field was not the only thing that was sinking in California. So were hopes for giant discoveries. After setting a production record in 1953 with an output of more than one million barrels per day, California's producers seemed unable either to push production higher or to sustain it at the million-barrel mark.

Wildcatters continued to find new oil and gas fields, but there were no more giants after San Ardo and Cuyama Valley. As the state's production dipped, there was growing pessimism that after more than sixty years of exploratory drilling, the biggest fields had been found.

Gloom set in, underscored by the feeling that time was running out for significant finds in traditional oil hunting grounds in the San Joaquin Valley, Los Angeles Basin and coastal basins. If wildcatters were going to hold the line on the state's oil production, it looked like they were going to have to broaden their horizons.

An obvious prospect area formed a horizon that for many Californians was the most beautiful vista in the state—the Pacific shore. For wildcatters, it was largely off limits. The State Lands Act of 1938 provided that only tidelands being drained of oil—or threatened with drainage—by wells on adjacent lands could be leased. The law shut the door on any search for new offshore fields.

Actually, offshore drilling was not new to California. In fact, it had been around for almost two decades longer than the Division of Oil and Gas. It was off California that the whole offshore business had started.

In the *First Annual Report of the State Oil and Gas Supervisor* that followed formation of the Division of Oil and Gas in 1915, Deputy Supervisor Robert B. Moran noted that the Summerland field in the Santa Barbara Channel, about five miles east of Santa Barbara, had produced about two million barrels of oil from 364 completed wells, the majority from 80 to 400 feet deep, located "very close together." A large number of the wells were drilled from piers extending out into the channel.

The field traced its beginnings to the purchase in 1883 of a portion of coastal land between the Carpinteria and Montecito Valleys by a man named H. L. Williams, who founded a colony he named Summerland for "spiritualists" who wished to live far away from society's commercialism. Four years later, Williams dug two water wells, both of which found oil. By 1895, Summerland had 28 oil wells. Ironically, the settlement meant to be a retreat became instead an oil town. The following year, in 1896, W. L. Watts of the California State Mining Bureau studied the region's geology and in his report wrote, "It is also evident that the oil yielding formations extend south into the ocean...At low tide, springs of oil and gas are uncovered on the seashore."

The first pier was built in 1897 to serve as a base from which to drill the nation's, and perhaps the world's, first offshore well. Other piers followed, including one 1,230 feet long. The piers bristled with stubby wooden derricks, making Summerland one of the scenic sights

along the route of the Southern Pacific's coastline, which the railroad liked to describe as the "Road of a Thousand Wonders."

In the *First Annual Report,* Deputy Supervisor Moran said drilling from the piers was "not such a difficult feat as might be supposed." Moran wrote, "The piers are light inexpensive affairs, owing to the fact that the channel waters are very quiet, there being almost no surf. The ocean water was excluded from the wells by driving casing into the clay beds overlying the oil sands. However, it is surprising that flooding has not been more rapid and complete than it was.

"Wells wrecked by a storm a number of years ago were never repaired. In 1912 broken casing could be seen under the water with small quantities of oil and gas continually escaping. The

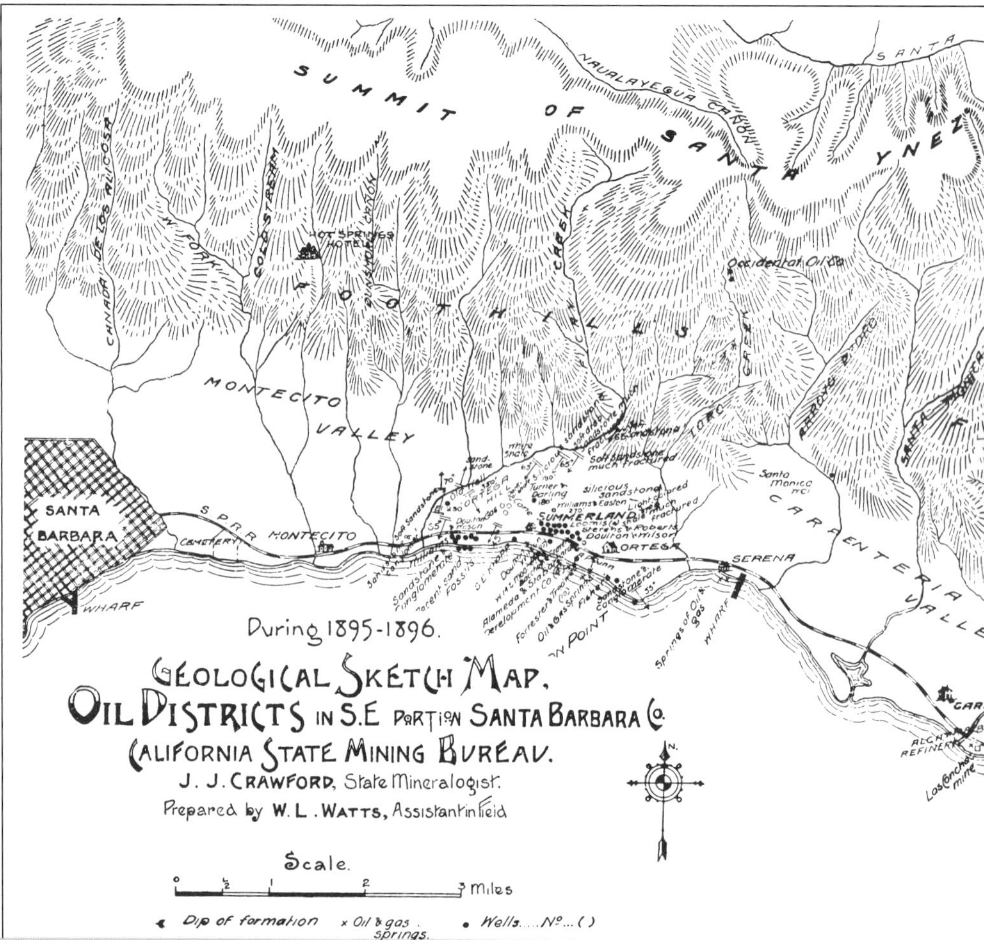

As early as 1896, W. L. Watts of the State Mining Bureau studied the geology of the onshore Summerland field and noted "... it is also evident that the oil yielding formations extend south into the ocean." The first pier was built in 1897 as a base for the nation's first offshore well. Above, Summerland circa 1904. (Long Beach Public Library.)

Summerland circa 1902. (Long Beach Public Library.)

field is also of note in that operations are continued, notwithstanding the very low production per well, which only amounts to 1.03 barrels per day. This is due partly to the fact that there is a local market for the product and also the fact that operating costs are very low. The wells are shallow and very close together and a large number can be operated from one power."

Production from the Summerland field continued until 1939 when S. G. Dolman, district deputy for District 3, in the *Twenty-Fifth Annual Report of the State Oil and Gas Supervisor* reported that high tides combined with storms destroyed the last shallow well in the ocean at Summerland.

Although many wells were drilled in the ocean from wooden piers at Summerland, it was not until 1921 that the State Legislature passed a definite tideland leasing act. Leases were good for a term of 20 years with five percent royalty going to the state. Many prospecting permits were issued for the offshore lands bordering Ventura and Santa Barbara Counties. Leases were issued to those who discovered oil.

A flurry of offshore activity came in the late 1920s following the discovery of the Rincon field eight miles northwest of Ventura by Pan American Petroleum Company in November 1927, and of the Elwood field 14 miles west of Santa Barbara by Barnsdall Oil Company in a joint effort with Rio Grande Oil Company in July 1928. Both discovery wells were drilled from onshore sites, but development soon extended into the tidelands.

In moving offshore at Elwood and Rincon, operators turned away from the wooden piers of Summerland's turn-of-the-century days to construct piers with steel pilings and reinforced concrete-filled caissons to support heavy drilling equipment. The caissons included a center conductor caisson and corner leg

Looking onshore, Summerland 1904. (Long Beach Public Library.)

With discovery of the Rincon and Elwood fields in the late 1920s, operators moved seaward again, pitching their tents on piers extending farther and farther into the Santa Barbara Channel. (Long Beach Public Library.)

supports. To conform with state requirements, rig floors were constructed as leak-proof aprons with a central sump designed to keep fluids out of the ocean.

As piers stretched as far as 2,300 feet into the ocean, the corner leg supports sometimes swayed under wave action. Operators began to use stabilizers—massive circular reinforced bodies of concrete—around the bottom of each corner column.

The Great Depression that began in 1929 and deepened in the early 1930s put a premium on finding ways to reduce costs, even if it meant going to sea without the security of a pier connected with the shore.

Indian Petroleum Corporation of Los Angeles failed to find commercial production with the first well drilled from an onshore site to explore its tidelands permit at Rincon, but the well did turn up geological information that pointed to probable production if a well were drilled further offshore.

In 1932, the company decided to forego the expense of building a costly pier in favor of simply building a portion of a pier, locating the structure at the site where geology appeared most favorable. The firm constructed a steel drilling platform in 38 feet of water some 1,700 feet beyond the end of the nearest pier. The structure quickly was dubbed the "steel island," and became California's first offshore drilling platform.

The operator drilled three wells from the platform. The small amount of oil the wells produced was piped ashore. In January 1940, mountainous waves battered the platform. The structure went down. There was no loss of life, but equipment was destroyed and wells damaged.

Beneath the floor of a drilling rig a surveyor takes a bearing to directionally drill a well, circa 1930s. (Long Beach Public Library.)

Rohl-Connolly Company, marine contractors, removed equipment,

Elwood oil field, 1932. (Long Beach Public Library.)

derrick and steel pilings from the ocean floor; cut off casing at the floor of the ocean; and placed 6-foot cement plugs in the tops of water strings.

In abandonment, C. C. Thoms, district deputy for District 2, in a summary of 1940 activities, noted that the steel island made one more bit of history. "It is the only instance we know of where the removal of oil well casing and placing of abandonment plugs was done by deep sea divers," Thoms wrote.

In Southern California, one aspect of the 1921 act that opened the tidelands was proving troublesome. It was the provision that no offshore prospecting permits or leases could be granted on lands fronting on municipalities and extending one mile on either side. Numerous cases of trespass occurred in the Huntington Beach field, though they did not come to light immediately.

In fact, production of oil from beneath the ocean in Huntington Beach field actually went on for some three years before it became generally known.

The unraveling of the Huntington Beach mystery began with the redrilling by The Superior Oil Company in May 1930 of a townlot well, Babbitt No. 1, and the recompletion of the well for a reported initial production of 346 barrels per day of 26-gravity oil. In April 1933, the production increased to 550 barrels per day and in August of that

The Great Depression that began in 1929 put a premium on finding ways to reduce costs. A Los Angeles operator chose to forego the expense of building a costly pier in favor of simply building a portion of the pier. The "steel island" at Rincon was the first California offshore drilling platform. (Los Angeles Public Library.)

The development of new tools made it possible to tap offshore oil from directionally drilled onshore wells, proving up new production in the Huntington Beach field. (Spence Air Photos.)

same year the output reportedly was 1,450 barrels daily. This was considered unusually good production for the Huntington Beach townlot area.

Meanwhile, Wilshire Oil Company drilled a well designated H.B. No. 15 on the same section and completed it in July 1933, flowing 4,815 barrels per day of 26-gravity oil. The horizontal drift, it developed, was in excess of 1,400 feet offshore from Ocean Avenue.

Before completion of the Wilshire well, many interested parties—including the Division of Oil and Gas—suspected that certain other wells were bottomed under the ocean, but very few had definite facts, and what knowledge they had was kept secret. However, the bottomhole location of Wilshire's well became known to several operators, and an intensive program of directing wells beneath the tidelands started immediately.

Shortly after this activity began, when it was certain that several wells had trespassed on state-owned tidelands, the Division of Oil and Gas and the State Lands Commission actively assisted the state attorney general to bring an injunction against one of the suspected operators as a test case, and to order a directional survey of the well.

The survey showed the well to be trespassing under state tidelands. Legal proceedings were instituted immediately against the operator and other suspected violators. This brought about a general slump in drilling activity. In November 1933, Governor James Rolph called a meeting of all interested parties. The next day, November 16, the Director of Finance declared a deadline after which operations could not start if the operator wished to retain the right to make an agreement with the State Lands Commission covering tideland operations.

By October 1934, sliding-scale royalty agreements, starting with a minimum of 5.42 percent for a 100 barrel-per-day rate for oil worth 50 cents a barrel and increasing with greater production or price, were made between the State Lands Commission and all operators. By the end of the year, about 85 wells were believed to be producing from the main zone offshore, with a total production for the year of 7,378,000 barrels, representing a rate of about 20,200 barrels per day. The Division of Oil and Gas determined Superior's Babbitt No. 1 to be the discovery well for the offshore pool.

With the passage of the State Lands Act in 1938, a new and far more orderly stage of offshore development began. Marine Exploration Company, which later became Monterey Oil Company, had a state tidelands permit off Seal Beach. The City of Seal Beach had an antidrilling ordinance that compelled the company to drill from outside the city limits. The first two wells were unsuccessful. However, their findings made it look like there might be something farther out.

Man-made Monterey Island served as a base in the 1950s for the development of Belmont Offshore field, one and one-half miles offshore from Seal Beach. (Monterey Oil Company.)

In 1948, Marine Exploration, drilling from an onshore site in Long Beach one-half mile inland from the water's edge, succeeded in whipstocking a well almost two miles seaward—9,271 feet— from the derrick floor. The drilled depth was 12,180 feet; the vertical depth was approximately 5,700 feet. The hole encountered oil sand. The company attempted to complete the well on pump, but was only able to put the pump 7,000 feet down the hole, leaving it 5,000 feet off bottom at a vertical depth of about 3,200 feet. The well produced a consistent 30 barrels per day of clean 26-gravity oil.

Encouraged, Monterey, successor to Marine Exploration, decided to build an island from which to develop the indicated field. The City of Seal Beach claimed a boundary three miles out to sea. Monterey decided it had two choices: seek repeal of the city's antidrilling ordinance or start operations and let the courts decide whether the city or state had jurisdiction over the tidelands. The company started building Monterey Island in 1952 at a site one and one-half miles offshore from Seal Beach in 42 feet of water. The city brought a criminal action. The case traveled through the courts for 14 months, with a final ruling by the California Supreme Court that the city had no jurisdiction.

Monterey, with The Texas Company as its partner, completed Monterey Island. Designed for 70 wells, the circular island, 75 feet in diameter, had an outer rim formed of interlocking sheet-steel piling driven into the ocean floor to depths of 15 to 20 feet. The interior was filled with rock and sand barged in from Catalina Island.

While Monterey was building the island, Brown Drilling Company, the contractor selected to handle the drilling work, marked out the exact dimensions of Monterey Island in the Lacey Trucking Company yard on Signal Hill. The contractor assembled within the outline a specially unitized National 50-A rig to be used for the drilling assignment, then moved the rig to an onshore site and drilled a well with the outfit set up exactly as it would be on the island. Drilling crews were required to stay in the "circle."

The first well on Monterey Island, itself, was spudded in May 1954. Some four months later it came in flowing 300 barrels per day from an interval at 5,906 to 6,280 feet. The field was

Standard Oil Company of California's Western Explorer *was an ex-Navy LSM (Landing Ship Medium) first converted to service as a lumber carrier and later as an offshore drilling vessel with the capability to drill in about 300 feet of water. (Division of Oil and Gas.)*

The first center-well drilling vessels mounted small rigs normally rated for depths of 2,500 to 3,000 feet at onshore sites. The vessels opened a new dimension in offshore drilling. Pictured is Richfield Oil Corporation's Rincon.*(Atlantic Richfield Company.)*

named Belmont Offshore.

After several wells had been completed, an upper deck was constructed on the island to house the drilling mast, making it possible to continue drilling while production equipment was being installed on completed wells on the lower deck. Before the end of the decade, 41 wells had been drilled from the island and some five million barrels of oil piped ashore.

So far, operators had managed to tap California's offshore oil from onshore sites, from piers extending almost one-half mile into the sea, from a steel drilling platform and from Monterey Island (a man-made island). But all offshore development still had one common link: it followed discoveries made from onshore sites.

While wildcatters could not seek offshore fields, they were allowed to make seismic surveys and do other preliminary work, including the taking of cores, to identify prospective producing structures for the day when they might be able to explore them.

The work called for the development of specially equipped vessels capable of drilling in the open sea. In 1950, Union Oil Company of California outfitted a former Navy patrol boat with a small rotary drilling rig for over-the-side drilling and rechristened the vessel *Submarex*.

During the next few years, several other ex-Navy boats were converted to over-the-side drilling, but they were only capable of drilling to shallow depths. In the mid-1950s, Standard Oil Company of California modified an ex-Navy LSM (Landing Ship Medium) to permit drilling through a 10-foot diameter hole cut through the hull in the center of the vessel. The company's *Western Explorer* had the capability to drill in about 300 feet of water. At about the same time, Richfield Oil Corporation acquired a similar hull and modified it for center-well drilling as the *Rincon*.

In the fall of 1956, another center-drilling vessel made its debut. The vessel was the *CUSS I*, the largest floating drilling vessel in the world. It was the outgrowth of three years of planning and work by an engineering group formed by Continental Oil Company, Union Oil Company of California, Shell Oil Company and Superior Oil Company.

The *CUSS I* represented an expenditure of about $3 million for the conversion of an ex-Navy YFNB

The queen of the offshore fleet was CUSS I, *a converted Navy freight barge that proved full-scale drilling was practical from a floating vessel. (William Rintoul.)*

Shell's Eureka *represented a new class of core drilling ships. The* Eureka *was equipped with a system that enabled the ship to hold a fixed position at sea without use of anchors and despite water depth. An electronic computer, which directed fore and aft propellers, formed the heart of the system. Control was automatic and accurate to within three percent of the water depth. The company used the vessel off the Pacific Coast to identify potential prospects. (Shell Oil Company.)*

freight barge into a vessel capable of drilling to depths of 10,000 feet or more in water depths up to 400 feet.

The powerful newcomer to California's fleet of drilling vessels utilized a diamond-shaped caisson through the hull with a modified National rig mounted well above the upper deck. The diesel power units and mud pumps were positioned on the main deck.

Innovative features included underwater television for observing the wellhead and an ingenious mechanical horizontal pipe-racking device that eliminated the need for a derrickman. Pipe was pulled and racked in doubles, that is, two joints at a time, by a system of elevators and chain hoists. The pipe was racked in a slotted conveyor system.

In the months that followed, the *CUSS I* proved that full-scale drilling was practical from a floating vessel. It handled assignments in water depths up to 1,500 feet, drilling holes as deep as 6,200 feet. It maintained position in all kinds of weather, cutting some 300,000 feet of hole during its first year of operation.

Even as deep-water technology was being developed, the State Legislature, looking at the decline in production that had set in after the peak in 1953, began moving to open offshore state lands for exploratory drilling.

Offshore looked like the best answer to the state's sagging production. Joe B. Hudson, Monterey Oil Company geologist, said at a meeting of the American Association of Petroleum Geologists in Los Angeles that a prize of more than 10 billion barrels probably lay waiting to be discovered in offshore portions of Southern California's Santa Maria, Ventura and Los Angeles Basins. Hudson said offshore seismic and geologic exploration in the past decade had turned up "well over half a hundred structures" that had been "mapped in detail." He warned that a great part of the offshore oil might never be recovered in the foreseeable future unless state and industry leaders established laws, rules and regulations under which the offshore oil could be produced economically and competitively in world markets.

In Sacramento, Assemblyman Joseph C. Shell introduced four bills to open the offshore door. The passage of the Cunningham-Shell Tidelands Act of 1955 permitted the State Lands Commission to lease offshore lands without the necessity of proving drainage, opening for possible leasing a stretch of California coast from Oceano, near Pismo Beach in San Luis Obispo County, south to Newport Beach in Orange County. The act allowed the drilling of wells from piers, filled lands or any satisfactory type of fixed or floating platforms. It set a fixed royalty rate of 12-1/2

The $2 million Pacific Driller *was a barge on stilts, equipped with a standard 136-foot derrick capable of drilling to 16,000 feet in water depths up to 90 feet. Standard Oil Company used the vessel to drill the coreholes that proved up the Summerland Offshore field. (William Rintoul.)*

percent for exploratory leases and gave the commission permission to set a fixed royalty of 16-2/3 percent or a sliding-scale royalty, beginning with that figure on proved oil lands. In all cases, competitive cash bonus bidding was required.

The first lease was awarded in January 1957. It was a 5,500-acre exploratory lease offshore from the pioneer Summerland field. Standard Oil Company of California and Humble Oil & Refining Company won the tract at a competitive sale with a cash bonus of $7,250,000.

Some legislators opposed the Cunningham-Shell Tidelands Act on grounds the state would not get a proper share of revenue. Soon after the Summerland lease was awarded, the State Lands Commission at the request of the Legislature suspended further leasing until the Legislature could study the 1955 act and make any amendments believed necessary.

While the study was going on, wildcatters continued to drill expendable exploratory core holes on prospective tidelands acreage with as many as ten core boats working. In the *Forty-Third Annual Report of the State Oil and Gas Supervisor* covering activity in 1957, Cecil Barton, district deputy for District 3, reported that of 106 notices to drill new wells filed during the year in the district, 85 were for offshore core holes.

One of the vessels that participated in the drilling flurry was a large mobile drilling barge built in New Orleans and towed through the Panama Canal to California. The $2 million vessel was the *Pacific Driller*. Ralph G. Frame, senior oil and gas engineer with the Division, described the barge as "a rather massive unit weighing...more than two U.S. Navy destroyers."

It was a barge on stilts, equipped with a standard 136-foot derrick capable of drilling to 16,000 feet in water depths up to 90 feet. The stilts were eight large caissons, each six feet in diameter by 195 feet long, which could be lowered to the ocean floor. Two DeLong air jacks gripped each 6-foot caisson. The jacks were operated simultaneously by air pressure to lift the drilling platform any desired height above the water, usually from 20 to 40 feet, depending on wave action. The barge also had a self-propelled, tread-mounted tractor crane for use in loading or unloading drill pipe and casing or performing any other lifting operations. A conveyor trough carried cuttings and waste mud to special hopper tanks, which were loaded onto barges by crane to be carried ashore for disposal. There was a special antipollution membrane to eliminate contamination.

Monterey Oil Company, acting as operator for a group that included Humble and Seaboard Oil Company, put the rig to work off Huntington Beach late in 1956.

From Platform Hazel three miles offshore from the original Summerland field, Standard Oil Company in 1958 began development of the first oil discovery made from the sea itself in California, opening a new era for the state. (William Rintoul.)

After the Monterey group finished its drilling program in late February 1957, Standard as operator for itself and Humble moved the *Pacific Driller* to the Santa Barbara Channel to drill a pair of core holes on the tract that had been leased from the state in January of that same year. The core holes found oil sand.

While Standard drew up plans to develop the tract, Richfield Oil Corporation at Rincon, eight miles to the southeast in the channel, moved one-half mile offshore into 45 feet of water to build California's second offshore drilling island to develop the seaward extension of the Rincon field.

The outer edge of the $4 million one-acre Rincon Island consisted of huge boulders. Successive layers toward the center became smaller and smaller, and the inner core was filled with sand. On the seaward side, the island was protected by 1,100 concrete tetrapods, each weighing 31 tons. The tetrapods, shaped like the jacks of childhood play, were built at a Carpinteria work yard and barged to the island.

The island was connected to the shore by a 3,000-foot causeway only wide enough for one-way traffic. The causeway curved up toward the middle to get above wave action so the two ends were not visible to each other. Ralph Frame described the system that was required when Division personnel and others traveled to and from the island. "In order to avoid the possibility of one vehicle meeting another on the causeway and one of them having to back up some 1,500 feet or less," Frame wrote in the *Forty-Sixth Annual Report*, "an electrically operated set of signals has been installed. A traffic light which flashes either green or red is set at each end of the causeway and when the driver of a vehicle approaches either end and sees the green light on, he knows the way is clear. However, before starting across the causeway he reaches through the car window and turns a switch, thus making the light show red on both ends until he has crossed the causeway and switched the lights back to green again."

Before 1957 ended, the State Legislature had completed its review of the Cunningham-Shell Tidelands Act and added amendments to change bidding and royalty provisions. The amended act stipulated that the State Lands Commission on both proved and unproved leases had to specify a sliding-scale royalty on oil production, commencing at not less than 16-2/3 percent up to a maximum specified in the bid. It allowed the commission to offer leases for the highest cash bonus or royalty.

In June 1958, the State Lands Commission held its first sale of tidelands leases following the amendment of the act. The sale ended a 17-month freeze and made available five tracts checkerboarded in the offshore area extending westerly from the Elwood oil field to Point Conception. High bidders offered more than $55 million in bonuses for the tracts, sending a strong signal that they were eager to look for more offshore oil.

In that same month, Standard towed an imposing tower a distance of 210 miles at a speed of 3 knots from the National Steel & Shipbuilding Company yard at San Diego to the Summerland tract. The tower was 75

Portions of Platform Hazel 40 feet below sea level. (Charles H. Turner, 1960.)

feet square and 170 feet high. It was a major component of Platform Hazel, and was to serve as the foundation on which a 110-foot square deck would be mounted to begin development of the tract. The tower was floated to the job site on the four big caissons that formed the bottom portion of the tower's legs, each 40 feet high and 27 feet in diameter. Each caisson was pressurized to prevent leakage and also ballasted with 90 tons of sand for stability.

To position the tower, Standard used a derrick barge with a 250-ton capacity boom, which steadied the tower and regulated the structure's descent as the buoyancy in the caissons was reduced. Once on bottom, the caissons were sunk 22 feet into the ocean floor by means of high-pressure water and air jets that literally hosed away the bottom sands, allowing the caissons to rest on hard ground. The final anchoring was accomplished by filling the caissons with 6,000 tons of sand and concrete.

After the tower had been placed on the ocean floor, the deck was barged to Summerland and installed. Designed for 25 wells, the deck stood 50 feet above the water, safely out of reach from storm wave action. On the deck was a one-of-a-kind, 162-foot Lee C. Moore derrick designed so two wells could be drilled at the same time. A sign of the times was a helicopter landing platform, which furnished a quick method for transporting personnel and supplies. The cost of building and installing the platform was $4 million.

On September 20, 1958, Standard spudded in with a Western Offshore Drilling & Exploration Company rig out of Long Beach to drill Standard-Humble Summerland State No. 1, using electric power transmitted from shore by a 16,500-volt submarine cable to run the National 80-B rig. The first well bottomed at 7,531 feet. On November 18, 1958, the well flowed 865 barrels per day of 36-gravity oil, proving up the Summerland Offshore field.

The discovery represented a significant breakthrough. Before, the move into California's tidelands had been in pursuit of discoveries made from dry land. The find at Summerland Offshore was the first new-field discovery made from the sea, itself, in California. A new era had begun for offshore development in the state.

Silhouetted against the setting sun, an offshore platform develops California's tidelands oil resource. The platform is Union Oil Company's Platform Eva, positioned in the mid-1960s to develop the offshore portion of the Huntington Beach field. (Stephen Mulqueen.)

9 Full Steam Ahead

In the early 1960s, rumors spread that steam was turning sluggish Kern River wells into big producers. In time, companies confirmed that steam was a potent new technique to recover hard-to-produce heavy oil. (William Rintoul.)

In the spring of 1960, Shell Oil Company quietly embarked on a new approach in an effort to relieve a frustrating situation that had confronted producers from the infancy of the California oil industry.

The dilemma involved the production of the heavy crude oil found in many of the state's fields. The crude was thick and sticky and all too often moved like molasses on a cold day. The positive side was that wildcatters had found billions of barrels of the viscous crude, by some estimates as much as 40 billion barrels. The negative side was that producing techniques were sadly lacking. It looked like operators would leave most of the oil in the ground. By most accounts, they would be able to recover little more than 10 percent. Shell chose the Yorba Linda field in the Los Angeles Basin to test a technique to recover more oil. The company began injecting steam into the Upper Conglomerate pay zone that produced 12-gravity oil from a depth of about 600 feet.

The theory was that the steam would lower the oil viscosity by heating the reservoir so oil would flow more readily, making it possible to pump greater amounts from the ground.

Scientists had tested the method in the laboratory, and results had been encouraging. With only a relatively moderate increase in temperature, heavy crude became much more cooperative, behaving like butter in a frying pan.

In a brief acknowledgment of the work at Yorba Linda, a Shell spokesman said it might be a considerable time before results from the pilot were known. If the pilot program proved successful, a full-scale operation might be attempted, but such an operation would be at least several years in the future. The company estimated the cost of the field test at up to $150,000.

Shell's approach was only one of various techniques companies were trying to apply heat to California's lazy oil. Four years before Shell began its pilot test, two fields in the San Joaquin Valley had become the setting for a method that seemed as direct an approach as one might conceive. In the South Belridge and Midway-Sunset fields, operators set producing sands on fire to heat the oil.

To allay fears that burning oil sand in the ground would result in most of the crude oil being consumed by combustion, researchers pointed to laboratory results that indicated — correctly, it developed — that no more than 10 percent of the oil would be burned, and that would be the least usable part, consisting of coke and heavy residuum. The lighter usable oil would be pushed ahead of the burning front to producing wells.

In the South Belridge field, General Petroleum Corporation picked the Marina lease for its pilot test. The sand selected for the experiment was an approximately 30-foot thick section at about 700 feet that occurred in the upper portion of the Tulare zone. The company utilized a five-well pattern. Four producing wells 330 feet apart formed the corners of a square with an injection well in the center through which air could be injected to support combustion.

Engineers designed an elaborate tool to ignite the oil sand by electrical impulse. The tool was so impressive they patented it. To set the stage for use of the tool, they began injecting air to build up a subterranean mixture that would support combustion. However, the oxidation of oil in the reservoir generated heat, and spontaneous ignition occurred

Heavy 12.6-gravity crude oil from Kern River, photo right, and lighter 25-gravity oil, photo left. (Division of Oil and Gas.)

before the tool could be used. Once burning started, it was necessary to continue injecting air into the oil sand to keep combustion going.

Ten other companies shared the estimated $1 million cost of the pilot test, including Continental Oil Company, Esso Research & Engineering Company, Gulf Oil Corporation, Honolulu Oil Corporation, The Ohio Oil Company, Shell Development Company, Sunray Mid-Continent Oil Company, The Texas Company, Tidewater Oil Company and Union Oil Company of California.

In the Midway-Sunset field, Standard Oil Company of California's subsidiary, California Research Corporation, selected fee land two miles east of Maricopa for a test burn at a depth of about 2,000 feet. The company drilled four wells, including an injection well and three producing wells, which encircled the injection well at distances of about 200 feet. Estimated cost of the project was approximately $1 million.

When combustion began on the hot July day in 1956, a potted plant with florist's foil wrapped around the container added a homey touch to the doghouse that doubled as an office and change room at the Midway-Sunset site. The container held a flame bush, a rare bush that grows in Southern California and produces a fiery bloom similar to a flame. It was a gift to Pete Simms, senior research engineer in charge of the project, from his wife and children.

Some public apprehension attended the initiation of underground combustion. There were some people who envisioned a raging fire bursting from the ground to engulf bystanders. Others expressed concern that the quantity of air being compressed for

The thermal recovery method tested by General Petroleum Corporation at South Belridge seemed as direct an approach as one might conceive. The company set producing sand on fire beneath the Marina lease, utilizing four producing wells, marked by pumping units, and an injection well in the center, marked by a standard steel derrick. (Mobil Corporation.)

In the Midway-Sunset field, Standard Oil Company of California picked a fee property two miles east of Maricopa for a fireflood experiment. The gin pole in the center of the picture marked the shallow well through which air was injected to support combustion some 2,000 feet below the surface. (William Rintoul.)

On a July day in 1956, John H. Thacher Jr., vice president of California Research Corporation, a subsidiary of Standard Oil Company of California, turned a knob on a small boxlike device to set sand on fire beneath his feet, initiating the Midway-Sunset field's first fireflood. (William Rintoul.)

From a doghouse command post, Joe Lindsay and Bob Dunlap, research assistants, kept close tabs on the control manifold that regulated the flow of compressed air and gas to the injection well at California Research Corporation's fireflood in the Midway-Sunset field. (William Rintoul.)

A fireflood begun by Mobil Oil Corporation, successor to the General Petroleum Corporation, in March 1960 on the Moco property in the Midway-Sunset field was believed to be the largest thermal-recovery project ever attempted. The burn was in multiple zones—the Monarch, Intermediate and Webster zones—at depths from 750 to 2,500 feet. Initially, six gas-fired compressors furnished air, including two 750-hp and four 1,000-hp units. In late 1964, the company began installing equipment to double air capability. (William Rintoul.)

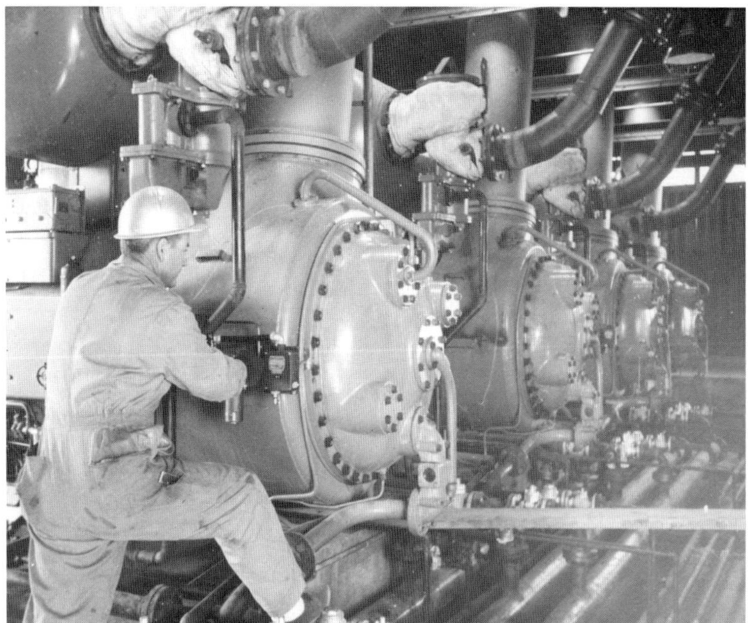

The equipment installed to increase capability for expansion of Mobil's Moco project at Midway-Sunset field consisted of nine free-piston engines, shown undergoing checkout. (William Rintoul.)

On the basis of such clues, engineers decided the rate at which air should be pumped downward. If air injection was too slow, the combustion front would come to a virtual standstill. If injection was too fast, the burning profile became distorted.

While fireflooders worked to perfect their technique, producers found another tool that enabled them to put the heat on heavy crude. The bottomhole heater applied heat opposite the producing sand. The system consisted of two separate units: a surface unit called the heater and a subsurface unit known as the heat exchanger. The surface unit came completely assembled on a metal base; the subsurface unit was assembled while tubing was being run into the well. Controls on the surface unit were regulated automatically by the pressure of the circulating pump and the temperature of the circulating fluid, which could be water or oil. The fluid normally was pumped into the well at temperatures up to about 300 degrees Fahrenheit and came back at about 200 degrees, making the trip to and from the heat exchanger in separate strings of small-diameter pipe run with the tubing in a single operation. The heat exchanger applied heat to the producing formation by radiation.

A strong selling point for the bottomhole heater was its cost. The $1 million outlay that seemed to be the entry fee for a try with the still-to-be-proved fireflooding technique effectively limited the ranks of fireflooders to major companies. On the other hand, for an investment of about $3,000, anyone

injection into the ground would result in a permanent low-pressure zone over test sites, causing perpetual storms.

As firefloods proceeded without undue incidents, fears subsided and engineers settled down to monitor the phenomena they had created in the earth. Various clues enabled them to keep track of what was happening beneath the ground. The temperature of oil coming from the reservoir indicated how close the combustion front was to the producing well. The composition of gases from below was another clue, because air injected into the oil reservoir was, like all air, composed of 21 parts oxygen and 79 parts nitrogen. If oxygen persisted in produced gases, the underground combustion obviously was inefficient.

could put a bottomhole heater to work. If the heater did not accomplish the desired results on a given well, there was always a chance it might do better on another.

The most successful use of bottomhole heaters came in the Kern River field, where Tidewater Oil Company installed heaters in 54 wells in 1957 and 1958, increasing production from 250 barrels per day to 1,443 barrels daily, a gain of almost 500 percent. The use of heaters by Tidewater and other operators largely was credited with boosting Kern River's production by some 2,000 barrels per day over a two-year period, from 12,390 barrels daily in 1956 to a high of 14,440 barrels per day in 1958.

While increased production at Kern River and in other heavy crude fields was a welcome turn of events, it was obvious to reservoir engineers that the heat radiating from heat exchangers penetrated only shallowly into the oil sand, representing something akin to trying to heat a huge room with matches. For better results, it was clear that heat would have to be delivered deeper into the sand.

Meanwhile results began to come in from the experiments with fireflooding. In the Midway-Sunset field, California Research Corporation discontinued its burn in May 1957 some 10 months after initiating combustion. The company did not release details, but the consensus among industry observers was that the project had not been commercial.

In the South Belridge field, General Petroleum Corporation had better luck, but there were still questions. The company confirmed that fireflooding had increased production from 20 barrels per day before ignition to a high of 390 barrels per day, averaging about 140 barrels daily for the four producing wells during the test, or seven times the rate before ignition.

With conventional methods of production, a spokesman said, a recovery of 10 to 15 percent of the oil in place might have been expected during the estimated 40- to 60-year life of the field. With fireflooding, recovery represented 51 percent over a one and one-half year period.

However, the cost per barrel of oil recovered was significantly higher for fireflooding than for normal production. The company reported it had been necessary to rework wells more often, including scratching perforations, swabbing, bailing and injecting hot oil into the casing. Corrosion and high temperatures had caused the failure of conventional steel pipe and equipment, both above and below the surface, creating costly maintenance problems.

The jury was still out.

In the months that followed the first injection of steam by Shell at Yorba Linda, there was no word from the company to indicate whether the pilot project was a success, or even if it showed promise.

However, before the year ended, Shell had begun preparations to inject steam in the Coalinga field. The company made no early announcement of the new field test, and in fact seemed disturbed when the news leaked out. The zone selected

Tidewater Oil Company engineers Ernie Young, left, and Mike Bealessio with one of the bottomhole heaters used to revive production in the declining Kern River field in the late 1950s. (William Rintoul.)

for the pilot test was the 1st zone, an isolated sand in the upper portion of the Temblor zone that produced 15-gravity oil from an average depth of 1,000 feet. The program called for drilling an injection well on Shell's Section 29 property around which were to be grouped four observation wells at a radius of 90 feet from the injector and two producing wells at a distance of 180 feet and on opposite sides of the injector. The observation wells were to be used for recording temperatures and to run frequent thermal surveys to detect any possible migration of steam from the oil zone.

In March 1961, Shell began injecting steam at Coalinga. Though the company had little to say about the new technique, the continued steaming at Yorba Linda and the expansion of steaming into the Coalinga field caught the attention of competitors.

In the Kern River field, Tidewater Oil Company, though enjoying more success than any other California producer with bottomhole heaters, had no illusions that the heaters would provide an ultimate, economic answer for large-scale application of heat.

Initially, the company's engineers favored hot water over steam, reasoning that if a limited amount

Portable steamers made it possible for operators to quickly—and relatively cheaply—see for themselves what steam could do. Above, two Halliburton portable steam-generating units on a job for Standard Oil Company of California in the Kern River field, June 1964. Ernie Gollehon was the Halliburton thermal operator. (William Rintoul.)

Steam generators proliferated almost overnight in the Kern River field, not only on major companies' properties but also on those of some independents. Above, a Crestmont Oil Company installation. (William Rintoul.)

of hot water in closed pipes in the bottom of a well, that is, a bottomhole heater, would bring a response in heavy oil production, a better response could be obtained by using more hot water, either pumping it directly into the producing well, or better, as a standard waterflood, using the hot water both to heat the oil and to push it through the reservoir to producing wells.

After laboratory work proved encouraging, the company began a two-well, hot-water injection test at Kern River in August 1961. The following year, Tidewater initiated a more ambitious heavy-oil recovery stimulation experiment that was code-named Project HORSE. The company began injecting water in August 1962, putting away the 300-degree water into four corner wells at a rate of about 2,000 barrels per day. Over a period of some eight months, the company injected almost two million barrels of hot water, increasing production from the center well to as much as 100 barrels per day, compared with an average production in the field of 7 barrels daily per well. Tidewater halted water injection when tracer tests indicated severe channeling.

While Project HORSE was under way, the company also began experimenting with steam, injecting into wells on an individual basis, allowing the well to soak and then returning it to production.

While there were differences in their approaches, both Tidewater and Shell shared one concern. For competitive reasons, each company operated with tight security.

Concurrently with steam, another interesting development began unfolding in California's heavy crude fields, especially those in the San Joaquin Valley. Almost overnight the producing leases owned by independent operators, mainly stripper leases, became hotly-sought-after properties in what before the end of 1962 was described as the greatest sellers' market in the history of California oil.

In the Kern River field alone, the buying spree resulted in more than 300 wells changing hands in a 10-month period. Prices rose to figures that were incomprehensible to many observers. Some sales brought prices as high as $9,000 per barrel of daily production. For example, if the property was producing 20 barrels per day, the price would be $180,000, an unheard of figure.

The prices made ancient history of those paid five years earlier, when $2,000 per daily barrel was considered the standard price, and even dwarfed those of only one year before, when $3,500 to $4,500 per daily barrel transferred producing properties from one owner to another.

The sudden sky-high prices were even harder to understand in view of what was happening at the same time with regard to the price paid for Kern River crude oil. While the price for producing properties was soaring, the barrel of Kern River crude that brought $2.55 five years before only brought $1.75 in 1962—a decline of more than 30 percent.

For the Division of Oil and Gas, the changes in ownership of leases and producing wells brought a

Chanslor-Western Oil & Development Company, which later became Santa Fe Energy Company, made an early turn to steam in the Midway-Sunset field. Above, CWOD's Duwain Rhoten inspected one of the company's generators in the mid-1960s. (William Rintoul.)

bumper crop of paper work in the form of map and well-record revisions.

Of more than passing interest was the fact that the two companies whose names were most often being substituted for the names of previous owners on maps and well documents were Shell Oil Company and Tidewater Oil Company, the leaders in experimenting with steam-recovery techniques.

In less than one year's time, Shell bought no fewer than 14 producing leases from six independent operators in the Kern River field, acquiring more than 200 wells. Those who sold to Shell included Nate Morrison, D. D . Lucas, E. A. Clampitt Oil Company of Oildale, P. D. Mitchell, Producers Oil Corporation of America and C. E. Rubbert.

Tidewater bought nine producing properties from five operators at Kern River, acquiring more than 70 wells from Ventura Oil Company, West Crude Oil Company, D & D Oil Company, Butler-Richardson and Producing Properties Inc.

While Kern River was the most active center for purchases, the same pattern was being repeated in other heavy crude fields. In addition to the Kern River wells, Shell purchased more than 300 wells in other San Joaquin Valley fields, including the Ralph R. Whitehall and McGreghar Land Company wells at Mount Poso; Pacific Coast Gasoline

Steam generator capable of producing 18-1/2 million Btu's in steam per hour was lifted off a "low-boy" trailer for an expansion of Tidewater Oil Company's thermal recovery operations in the Kern River field. The portable generator was placed on a timber- and gravel-pad to produce steam for injection into the shallow Kern River sand. (Tidewater Oil Company.)

Engineers Lawrence and Hutchinson at one of thirty steam-liquid separators included in the casing-blow system installed by Tidewater Oil Company in the beginning days of large-scale steaming at Kern River. (Tidewater Oil Company.)

Oil-fed fires in steam generators converted water to steam to heat Kern River oil sands in the beginning days of the Tidewater Oil Company steaming operations. (Tidewater Oil Company.)

Company and Anchor Oil Company leases at Midway-Sunset; Maxwell Hardware Company, Oil Division, leases at Round Mountain; Producers Oil Corporation of America lease at Coalinga; and Producing Properties Inc. leases in the Edison, Mountain View, Mount Poso, Round Mountain and Coalinga fields.

Though no information was being released, it was abundantly obvious that there had been a breakthrough in thermal recovery technology that would add value not previously present to heavy-crude producing properties.

Overnight, there was a scramble to get a part of the steam action. The acquisition of leases at unheard of prices by Shell and Tidewater had to mean that the economics looked good. And unlike fireflooding, the outlay to become an oilfield steamer was not so large that it eliminated the small operator automatically. Steam not only seemed within reach for most producers, but also seemed to promise an almost immediate return. With luck, it looked like an operator might invest $30,000 for a steam project and make enough in six months to pay off the investment.

There was no comparison between the investment required for steam and that for fireflooding. A steam generator capable of converting 1,200 barrels per day of water into steam could be purchased for $35,000 to $45,000. A smaller generator capable of converting 500 barrels per day of water into steam could be bought for about $20,000. For fireflooding, an operator was looking at an outlay of more than $100,000 just for a compressor capable of furnishing the air necessary to sustain underground combustion, and the return would not be so fast. Time was required for the heat from combustion to make itself felt in increased production.

Steaming had another immediate advantage. The interested operator could rent a steam generator for a pilot test before making any substantial investment in steam generators. Soon the waiting list for steam generators ran three months or longer. Manufacturers of both steam-generating units and water-treatment equipment enjoyed boom times. In four months one salesman reportedly sold more than $1 million worth of steam generators.

In light of the suddenness with which steam captured California's heavy-crude fields, the competitiveness of the situation and the rapid development of the technique, most operators chose to say nothing about their steam projects. Most were reluctant even to acknowledge projects, much less release information on results.

To find out what competitors were doing, operators turned to a branch of the oil industry whose exploits most often were associated with learning what was happening at exploratory wells. Oil scouts accustomed to counting stands of pipe being pulled from "no dope" wells to learn how deep they were and to looking for tell-tale clues like oil in the sumphole or casing on the rack to indicate someone had turned up shows found themselves cast in a new and unexpected role. Scouts traveled oilfield roads counting steam generators, noting who was using them and trying to find out whether they were getting encouragement. Operators responded with "Keep Out" signs, locked gates and sometimes guards. One major company placed bags over pressure gauges. Rumors flew of unbelievable increases in production from shallow wells that once had seemed like candidates for early abandonment.

From Kern River came word that Tidewater Oil Company in a field test of an undisclosed number of shallow wells had increased production from a pre-steam rate of 100 barrels per day to a post-steam rate of 430 barrels daily.

In the Midway-Sunset field, Mobil Oil Corporation, the successor to General Petroleum Corporation, was said to have boosted output from a well that had been pumping 10 barrels per day to a post-steam rate in excess of 100 barrels per day.

At South Belridge, it was rumored that Belridge Oil Company had injected steam for 15 days into a well that had been pumping 27 barrels per day and returned the well to production making more than 250 barrels per day.

In the trial-and-error beginning, the most popular method of steaming was what field men called huff-and-puff. The technique involved injecting steam into a producing well for a given period, perhaps 10 days; allowing the well to sit, or "soak," for perhaps another 10 days; then returning the well to production.

A logical next step was steam-drive, or steamflooding, in which steam was injected continuously into an injection well to heat and push oil toward producing wells.

With the spread of steaming, the secrecy surrounding the work began to fall away. In its first July issue in 1964, *California Oil World*, now a twice-monthly publication, reported, "The magic word is 'steam' as far as assisted recovery in California is concerned. Even as wildcatters look to the state's offshore prospects for the big discoveries of tomorrow, engineers concerned with increasing recovery from proven fields are turning to steam as a potent means of getting out more barrels quicker and more efficiently."

The article included a list of active projects, qualifying the list with the statement, "If the list seems incomplete, it might be well to keep in mind most operators are reluctant to even acknowledge their projects, much less to describe them—or the results."

The list included 41 projects in the South Belridge, Casmalia, Cat Canyon, Coalinga, Cymric, Edison, Guadalupe, Kern Front, Kern River, Lynch Canyon, Midway-Sunset, Mount Poso, Paris Valley, San Ardo and Yorba Linda fields.

The magazine's second January 1965 issue reported, "If the past year is remembered for nothing else, it will go down as the year production engineers 'discovered' steam."

Recounting how drillers once had used steam to power rigs and pipeliners had employed it to heat crude for shipment in pipelines, the article continued, "Neither drillers nor pipeliners, however, tapped the full potential of steam. It remained for the production engineer, particularly the man with an eye on the barrel nobody could quite produce, to come up with the ultimate weapon. And as a result, California's oil fields, like the New World in the wake of Columbus' discovery, aren't ever going to be quite the same."

The *Fifty-Second Annual Report of the State Oil and Gas Supervisor* covering activity for the year 1966 inaugurated a new feature that underscored the significance

Bank of steam generators, Kern River oilfield, with a Chevron USA Inc. injection well in the foreground. (D. Isenhower.)

of what was happening in the state's fields.

A table titled "Steam and Hot Water Injection Statistics" listed cyclic and flood projects in a field-by-field tally that included a breakdown of pools being stimulated. The total for the state showed that during the year operators had injected 110.6 million barrels of water converted to steam into 6,753 wells, or one out of every six of the state's 41,334 producing wells.

The greatest activity was in District 4, where operators injected 76.9 million barrels into 5,547 wells, representing 70 percent of the water injected as steam and four out of five of the steam-stimulated wells. The most active fields for steam injection were Kern River, 44.3 million barrels; Midway-Sunset, 18.2 million barrels; and South Belridge, 5.7 million barrels.

The following year, California's oil production topped one million barrels per day for the first time since 1953. A year later, State Oil and Gas Supervisor Fred Kasline reported that California had produced 373.2 million barrels of oil in 1968, establishing a new production record by topping the 367.3 million barrels produced in 1953. Steam was credited with a major role in the resurgence of production. Through the years following August 1961 when the first Kern River project began, the field's production increased each year—with the exception of 1970 and 1974— until 1986 when crude-oil prices collapsed. In 1966, the field topped the pre-steam record of 17.2 million barrels set in 1904 and in 1986 produced a record 47.8 million barrels, or almost six times as much oil as the 8.6 million barrels in 1961 when the first steam project was tried.

In the Midway-Sunset field, the first steam project began in October 1963. Production increased each year for the next five years, declined slightly in 1969 and 1970, then continued to increase for the next 17 years through 1988. In 1972, the field topped the 1914 production record of 34.4 million barrels. In 1988, the field produced 58 million barrels, an increase of three and one-half times over the 16.6 million barrels produced in 1963 when steaming began.

The South Belridge field had its first steam project in February 1963. That year the field produced 7.5 million barrels, a new record easily topping the pre-steam level of 4.6 million barrels set in 1945. With the exception of five scattered years, production increased each year through 1987, when the field produced 63.6 million barrels, or eight and one-half times more than when steaming started. Though production in the South Belridge field declined 3.2 million barrels in 1988, the field's showing still was strong enough to make it the third most productive field in the United States, runner-up only to Alaska's Prudhoe Bay field with 576.3 million barrels and the same state's Kuparuk River field with 112.1 million barrels.

The Midway-Sunset field was the fourth most productive field in the United States with 58 million barrels and the Kern River field the fifth most productive with 47.3 million barrels. In the three San Joaquin Valley fields in 1988, operators put away 1.3 million barrels per day of water converted to steam, firmly establishing the three fields as the thermally enhanced oil-recovery capitols of the nation's oil fields.

10 Tapping Nature's Teakettle

One half of a stereograph titled the Devil's Kitchen, *taken around the turn of the century by Andrew Price, in the days when The Geysers was a famed health resort. (Library of Congress.)*

While some busily put steam into the ground, others concerned themselves with taking it out. The focus for the effort to harness the earth's geothermal resources was a remote Northern California canyon, 75 miles north of San Francisco, which with its hissing columns of steam and bubbling mudholes had been described by a member of John C. Fremont's survey party in 1847 as "the Gates of Hell."

William B. Elliott had come upon Big Sulphur Canyon in the Mayacmas Mountains of northern Sonoma County while trailing a marauding grizzly bear. Stepping over a rise he had been surprised by clouds of steam issuing from roiling fumaroles. When he had returned to the Fremont party in Kelseyville, he had described the place as The Geysers.

Apparently there had been no one within earshot to explain that a geyser is a spout of water and a fumarole is a vent of steam. The name stuck, although there are no geysers at The Geysers.

A road was built to The Geysers in 1863, and a resort followed. For many years, visitors to Northern California were told their visit was incomplete without a trip to The Geysers Resort and mineral hot springs. As a place to bask in sulphur baths and exclaim over the hillside that spouted steam, The Geysers attracted the wealthy and famous, among them J. Pierpont Morgan, Ulysses S. Grant, Horace Greeley, Theodore Roosevelt, William Jennings Bryan, Mark Twain, Jack London, Luther Burbank and the Prince of Wales before he became King Edward VII.

Among those who came to The Geysers to admire the view and take the "baths" were some who saw in the billowing steam a source for power production.

One such visionary was John D. Grant, owner of a gravel pit in Healdsburg, who in the summer of 1921 began drilling on the hillside east of Big Sulphur Creek in the hope of harnessing steam for the generation of electricity. Impressed by the escape of steam at the surface and its relatively high temperature, Grant theorized that both volume and temperature would increase with depth.

At the time, Grant was unaware that a similar project had been successfully undertaken a number of years before at Larderello in Tuscany, Italy, where hissing fumaroles, bubbling pools and sulfurous odors were said to have inspired Dante's *Inferno*. In 1904, electric bulbs had been lighted at Larderello by a small generator driven by a steam engine run by natural steam flowing from a fissure in the ground. The next year, the world's first geothermal power plant was built at Larderello, furnishing electricity that, among other things, helped propel Italy's electric trains.

Grant's drilling rig was a homemade cable-tool outfit put together in Healdsburg. The first well quickly proved his contention that the volume and temperature of steam would increase with depth. At a still shallow depth, the well "blew up like a volcano," recalled Glen Truitt, who worked in the drilling crew.

The following summer, Grant drilled another well on the eastern bank of Big Sulphur Creek. After penetrating a sandstone cap at 80 feet, the drilling crew ran an 8-inch steel casing and "cemented" the casing with several hundred pounds of molten zinc. Drilling ahead in open hole, they encountered steam just below 200 feet. "Everything came flying up," Truitt recounted. "Mud, tools, rocks and steam. After things settled down, there was just clean steam. But the noise was loud enough to hear all over the valley."

The crew closed a heavy gate valve at the top of the casing, shutting in the first geothermal well in the western hemisphere.

Grant obtained an aging steam engine from the Foppiano Winery in Healdsburg and rigged it up to allow him to use steam from the successful well to drill another well within 50 feet of the location. The well was drilled to a depth of 318 feet and successfully completed in July 1923 in the same manner as the first. Grant contracted to sell electricity to the City of Healdsburg.

A Sunday supplement writer, taking note of the project, concluded that Grant's steam wells could spawn other projects actually tapping the earth's molten core. "This cheap energy could glut the world with wealth," he wrote, envisioning an end to the "greeds, envies and wars engendered by materialism."

One power company official, watching the development closely, estimated that enough wells could be drilled to develop a total of 112 megawatts of generating capacity.

Andrew Price, photographer, recorded with stereographs the days around the turn of the century when The Geysers was a famed health resort. In the first stereograph, "Calistoga" is painted on the wagon beneath the driver's elbow, a suggestion the visitors might be arriving from that city. The idea seems to be confirmed by Robert Louis Stevenson, who wrote in 1883 in The Silverado Squatters, *that the railroad ended at Calistoga and "...the traveller who intends faring farther, to the Geysers or to the springs in Lake County, must cross the spurs of the mountain by stage." (Library of Congress.)*

In spite of the project's promise, costs and technical problems interfered with Grant's plans. A glut of oil hit the West Coast market, making electricity from oil-fired steam plants more economical. The Geysers project lapsed in the early 1930s, but not before generating enough electricity to light The Geysers Hotel and streets so that, in the words of Glen Truitt, they "looked like a Christmas post card."

Some 20 years later, two more entrepreneurs appeared on the scene. One was B. C. McCabe, the president of Magma Power Company, of Los Angeles. The other was Dan McMillan, president of Thermal Power Company, of San Francisco. The two were boyhood friends. McMillan had been the star tackle on coach Andy Smith's Golden Bears "Wonder Team" of 1920-21 at UC Berkeley. The powerhouse team had smashed Ohio State in the Rose Bowl, 28-0, capping a season in which the team scored 510 points and allowed opponents only 14 points.

In 1955, Magma Power Company leased 3,200 acres at The Geysers. The following year, Thermal

Power Company joined the venture. With $270,000 between them, McCabe and McMillan established the Magma-Thermal Power project, leased another 2,800 acres and began drilling with rotary tools.

In 1957, Magma-Thermal drilled or redrilled six wells to depths varying from 500 to 1,400 feet. A flow test of four of the wells proved that an economic supply of steam was available. If there was any doubt, it was dispelled by another of the wells that in September of that year blew out, possibly from a fault downhill of the well site. At first, drillers believed the well had tapped a live volcano. They called it "The Outlaw."

"When she blew," McMillan recalled, "more than 3,000 tons of rock and debris flew all over the countryside and the noise was deafening."

A crater swallowed up the drilling rig. Workmen tried to kill The Outlaw by flooding the hole with cement slurry, but the steam blew it out, raining wet cement over the valley. When more cement was poured into the hole, the steam squirted out from other holes in the hillside. Workmen finally succeeded in partially taming the wild well by installing pipe through which to capture a portion of the steam, but even today, after many more efforts, some steam continues to escape from the well.

Though Magma-Thermal had harnessed enough steam to power a generating plant, there were still problems to be solved before the geothermal resource could be translated into electricity. The steam carried minute fragments of rock, which, if not removed, would damage the turbine blades of the generating plant. Engineers devised centrifugal "cleaners" to spin out the particles by "twisting" the steam a few turns before it started its course by pipeline to the generating plant. Magma-Thermal installed cleaners at each wellhead.

One of the most difficult problems that had to be met in designing the plant was overcoming the corrosive characteristics of the natural steam. Tests showed that carbon steel was useless. Researchers found stainless steel alloys were capable of withstanding the corrosion at any point in the plant where steam would come in contact with metal.

These problems were overcome, and on September 25, 1960, Pacific Gas & Electric Company (PG&E) dedicated its Geysers Power Plant Unit 1, opening a new chapter in the history of power generation in the

Early Geysers' power plant in the early 1930s generated enough electricity to light The Geysers Hotel and street so they "looked like a Christmas post card." (Brian Roberts.)

United States. The plant was the first in the nation (since the attempt by John Grant) to harness geothermal energy for the generation of electricity, utilizing 240,000 pounds of steam per hour supplied by four Magma-Thermal wells to turn turbine blades generating 12.5 megawatts, said to be enough electricity to satisfy the requirements of a city of 12,500 people. With successful operation of the plant, PG&E could proudly claim the distinction of being the only utility in the United States using five sources to generate electricity, including falling water, fuel oil, natural gas, nuclear energy and, now, geothermal steam.

The campaign to develop steam production accelerated. In spite of subsurface temperatures of more than 400 degrees Fahrenheit and an ever-present danger of blowouts, drillers wearing ear muffs to protect their hearing from the thunderous roar of steam and blackboards and chalk with which to communicate, succeeded in completing ever deeper wells in the fractured rock "pay" that contained an apparently inexhaustible supply of steam created by heat from the earth's interior.

Some put steam into the ground to recover more oil. At The Geysers, others drilled wells to take steam out of the ground, as did Brown Drilling Company in the early development of the field. The cloud of steam coming from the muffler signals a successful test. (William Rintoul.)

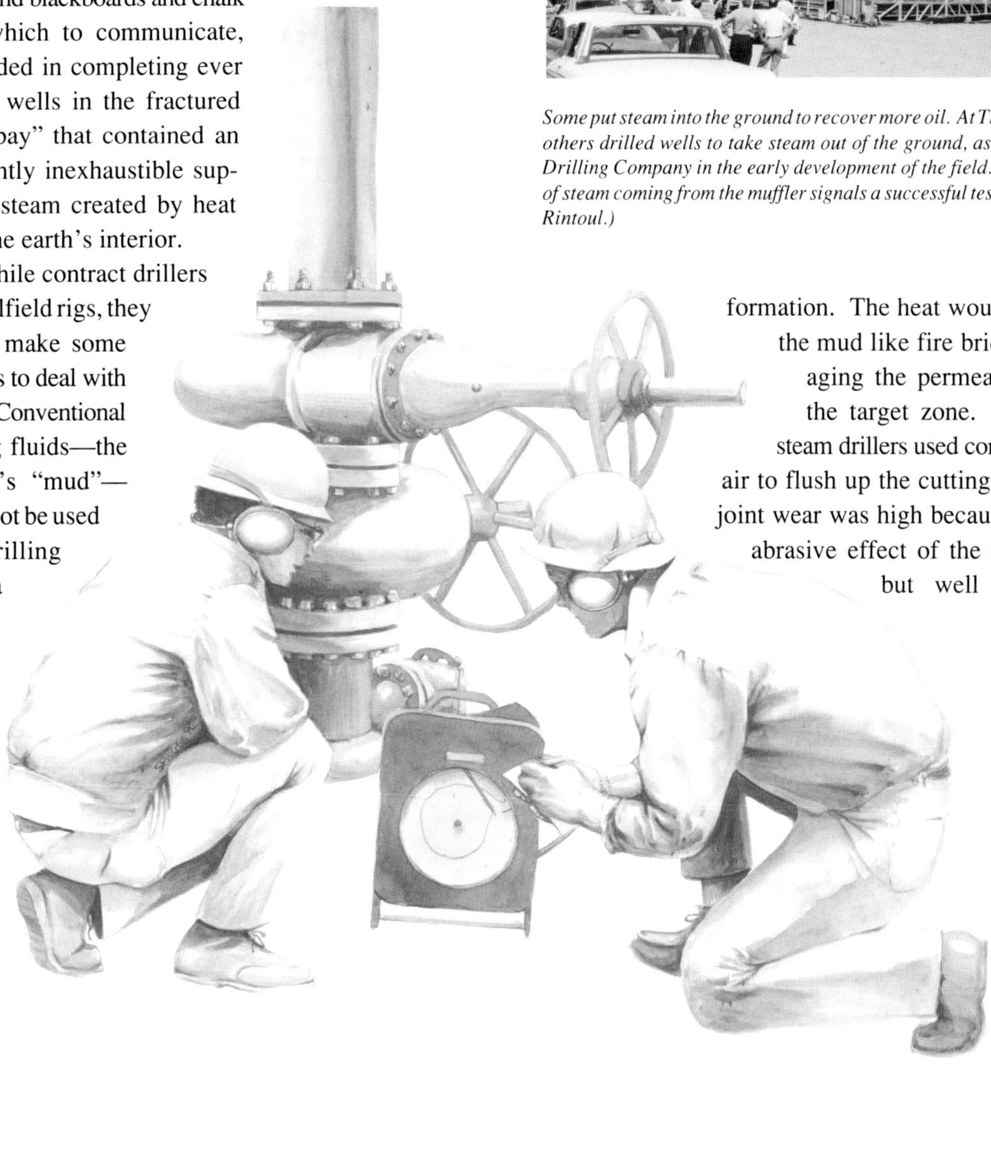

While contract drillers used oilfield rigs, they had to make some changes to deal with steam. Conventional drilling fluids—the oilman's "mud"—could not be used for drilling into a steam formation. The heat would set up the mud like fire brick, damaging the permeability of the target zone. Instead, steam drillers used compressed air to flush up the cuttings. Tool-joint wear was high because of the abrasive effect of the cuttings, but well cleanup proved relatively easy. The wells were allowed to discharge into the atmosphere to remove particles and keep the well bore clean.

Before the end of 1965, Thermal Power Company, acting as operator for itself and Magma Power Company, had completed 29 wells capable of producing dry steam. The wells offered enough capability not only to supply a second PG&E unit, which went on line in March 1963 utilizing 265,000 pounds per hour of steam to generate 14 megawatts, but also to spur the start of construction of a third plant, designed to use 580,000 pounds per hour of steam to generate 27.5 megawatts.

In that same year, the State Legislature enacted a bill that gave the Division of Oil and Gas, effective September 17, 1965, the authority to supervise the drilling, operation, maintenance and abandonment of wells drilled for the development of geothermal resources in a manner similar to its supervision of oil and gas wells.

The *Fifty-First Annual Report of the State Oil and Gas Supervisor*, covering 1965, recognized the new responsibility with the first of what would be an ongoing series of geothermal reports, summarizing each year's activity.

Another new player became an important part of the California geothermal scene in 1965. Union Oil Company of California became involved as a consequence of the company's merger in July with Pure Oil Company, which was active in geothermal exploration in the Imperial

A new player became an important part of the California geothermal scene in 1965 with the involvement of Union Oil Company of California in The Geysers field. Pacific Gas & Electric Company constructed a series of generating plants to utilize the harnessed steam. (Unocal Corporation.)

Pacific Gas & Electric Company Power Plant Units Nos. 7 and 8, at The Geysers geothermal field. (Pacific Gas & Electric Company.)

Valley. Dr. Carel Otte, who had directed Pure's geothermal effort as operating head of Earth Energy, Pure's geothermal subsidiary, joined Union. Otte was an Amsterdam-born geologist who, after wartime service as a captain in the RAF, had earned his doctorate at the California Institute of Technology. He had served at Pure's Crystal Lake Research Center until his appointment to Earth Energy in 1963.

Earth Energy, now a Union Oil subsidiary, made its debut at The Geysers with a well on the Ottoboni lease in June 1966. At 5,200 feet, the well came in with a cloud of steam. "It was a significant stepout," Otte said later, "more than the entire width of the Sulphur Bank producing area."

In early January 1967, Union's Earth Energy, Magma Power Company and Thermal Power Company signed a letter of intent to pool their Northern California lands for a combined holding exceeding 14,000 acres in The Geysers area, with the Union subsidiary to serve as operator.

At the time, steam from Magma and Thermal wells was being used to generate 26 megawatts. An additional plant, the third in The Geysers complex, went on line in March, adding 27.5 megawatts to more than double the previous capacity.

In announcing the agreement, Fred L. Hartley, Union's president, said, "This is Union's first move into the commercial use of a form of energy other than petroleum or natural gas. We are most optimistic about its potential as a means of producing dependable and economic electric power."

Hartley added that Union contemplated an aggressive program to develop the new natural resource.

While The Geysers was giving California and the nation their first commercial electricity from geothermal energy, wildcatters were targeting other parts of the state for geothermal development. The

Loffland Brothers Company drilling rig on location for Union Oil Company of California in The Geysers geothermal field. When President Gerald Ford and a group of administration officials visited the field in April 1975, a Loffland rig and other facilities were on the agenda for the Presidential party, which arrived in cool and cloudy weather and was drenched with rain, snow and hail before the two-hour visit was concluded. (Unocal Corporation.)

Whenever possible in developing The Geysers field, Union Oil Company of California located pipelines in secluded areas along natural routes away from public view. Here, an employee checks an insulated pipeline that carries geothermal steam from wells to generating units. (Unocal Corporation.)

While operators were developing The Geysers geothermal field, wildcatters were targeting the Imperial Valley for exploration and development. In April 1989, Unocal dedicated the $110 million power plant, Salton Sea Unit 3, in the midst of farm fields eight miles west of Calipatria at the southern tip of the Salton Sea. (William Rintoul.)

As a measure of the Salton Sea field's capability, Unocal initiated power generation at the 47.5-megawatt, net, Salton Sea Unit 3 plant with only two producing wells. One, the well Vonderahe No. 1, right foreground, was described by the company as the world's largest geothermal well, producing enough brine for 30 megawatts, or enough electricity to meet the needs of about 30,000 residential customers. (William Rintoul.)

Imperial Valley, like The Geysers, had attracted steam drillers in the 1920s. In fact, the search for geothermal steam in the Niland area near the Salton Sea had led to the discovery of commercial quantities of carbon dioxide in 1927. By the mid-1940s, more than 30 wells completed at an average depth of 250 feet were on stream, producing carbon dioxide that was used for making dry ice. The field subsequently was abandoned when the rising level of the Salton Sea flooded well sites, and other methods of refrigeration proved more competitive.

In the early 1960s, a wildcatter drilling for oil in the Niland area accidentally discovered a hot brine reservoir with temperatures ranging from 400- to 600-degrees Fahrenheit. By the end of 1964, at least 10 wells had been drilled to evaluate the resource, but the extremely high content of dissolved solids in the brine, running over 250,000 parts per million from some wells, posed a formidable stumbling block to any immediate attempt to use the fluid for the generation of electricity. The quantity of salts in the brine led to an attempt at recovering potash by both Morton International and Pure Oil Company. However, the collapse of the price of potash in 1963, together with corrosion problems, caused curtailment of that program. However, research with highly mineralized brine continued, and between 1982 and 1989, six geothermal power plants, all using crystallizer-clarifier technology to handle the heavy mineralization, were built in the Salton Sea geothermal field by subsidiaries of Unocal Corporation and Magma Power Company.

To the south in the Heber area, five miles south of El Centro, another sector had caught the eye of

Magma Power Company's J. J. Elmore Geothermal Power Plant at the Salton Sea field. The power plant began operating in December 1988, and has a nameplate rating of 38 megawatts. (Susan Hodgson.)

The dual-flash power plant in Heber geothermal field was dedicated on October 31, 1985, producing 47 megawatts, net, of electricity. (Susan Hodgson.)

geothermal prospectors as early as 1945 when an exploratory well drilling for oil turned up water temperatures so high that drilling fluids had to be cooled with dry ice. Amerada Petroleum Corporation's Timken No. 1 had been abandoned at a total depth of 7,323 feet, but the temperature shows had not been forgotten. Chevron USA Inc. began delineating the Heber reservoir in 1972 with an eye toward commercial development. Eventually, the first two large geothermal power plants built in the Imperial Valley, one a binary and one a flash plant, were constructed at Heber. Both were dedicated in 1985.

But it was earlier, in 1980, that two much smaller geothermal power plants were dedicated at Brawley and East Mesa geothermal fields. These were the first commercial, Imperial Valley geothermal power plants. Both were 10-megawatt power plants, again one a flash plant and one a binary. The projects were developed by Southern California Edison Company and Unocal Corporation (at Brawley), and Magma Power Company (at East Mesa).

In fact, it was the East Mesa geothermal field, some 17 miles east of El Centro and the site of the East Mesa heat anomaly, where the U. S. Bureau of Reclamation in the early 1970s drilled an 8,000-foot well as the first phase of a $16 million project to determine if it might be economical to use geothermal heat to desalinate water. Eventually, Magma Power Company, Ormat Energy Systems, Inc., and GEO East Mesa Limited Partnership built geothermal power plants there.

The Coso Hot Springs area in the Mojave Desert, included in the China Lake Naval Weapons Center north of Ridgecrest, also began to take shape as a possibility for significant geothermal development. The area's potential had been recognized as early as 1925 when H. N. Siegfried, an engineer for Southern Sierras Power Company, later to become a part of Southern California Edison Company, said of Coso, "There is no doubt that a survey will disclose ample land still free for entry that will be suitable for natural steam development purposes."

The establishment of the China Lake Naval Weapons Center in World War II had removed the land from entry for such purposes as geothermal development, but the idea had hardly died.

The notion that the Coso geothermal resource could be developed to further the Navy's mission at China Lake came to the fore in 1964 when Dr. Carl Austin, who had come to the base in 1961, published the first Navy report on the exploration for geothermal potential in the Coso area. Convinced that the resource existed and that it could be developed for the

At groundbreaking ceremonies in March 1986 for the first geothermal power plant in the Coso geothermal field, California Energy Company opened up one of what were then eight wells drilled in the field, blasting steam high into the air. (William Rintoul.)

Joshua trees and desert scenery provided a setting different from the forested Geysers area for Loffland Brothers Company's Rig No. 98, at work in the 1988-1989 drilling boom in the Coso geothermal field. (William Rintoul.)

benefit of the Navy, Austin tackled the job of convincing skeptics at the Naval Weapons Center that the resource was worth developing.

The opposition buttressed its position with a memorandum put out in February 1964 quoting a U.S. Geological Survey representative as saying that the steam had little or no commercial value and that other geologists at the Naval Weapons Center did not agree with "Doc Austin's conclusions."

Austin proved persevering and persuasive. In 1966, the Naval Weapons Center drilled its first shallow test well to obtain chemical, temperature and heat-flow data for the proposed geothermal resource area. Five years later, Austin published *Geothermal Science and Technology, A National Program*, which marked the formal beginning of the Navy's geothermal project at Coso. California Energy Company later signed on as the contract developer, and the campaign began to turn Coso's geothermal energy into electricity. By the end of 1989, the company had completed Phase I of its project with the construction of nine geothermal power plants.

At The Geysers, with Union Oil Company of California spearheading continued development and PG&E commissioning new power plant units to keep up with burgeoning steam production, a significant milestone was reached in 1973.

In December of that year, PG&E's generating complex at The Geysers became the largest geothermal power facility in the world. Completion of the plant's tenth generating unit raised capacity to 412.5 megawatts, topping the Larderello, Italy, development, which was rated at 405.6 megawatts. Though now the largest in the world, the complex at The Geysers remained the only commercial geothermal power development in the United States, and PG&E announced plans to build four more units in addition to one already under construction.

As geothermal exploratory and development activity increased, the Division of Oil and Gas played a role not only in supervising field operations, but also in publishing statistics and disseminating information helpful to those in the industry. In 1970, the Division began publishing the *Geothermal Hot Line* as a periodical covering significant developments in the geothermal industry. In addition, growing interest in geothermal energy was met with other public outreach measures. In one year alone—1972—the Division's Geothermal Unit headquartered in Sacramento, with a staff of two engineers, in addition to its regulatory duties attended and/or spoke at about 100 meetings, conferences, field trips and conventions.

Four years later in 1976, to provide operational and testing services in step with increased geothermal activity in California, the Division set up three geothermal district offices, including one to oversee most of Northern California (in Sacramento); one for all of Southern California (in El Centro); and one to

Three power plant units in operation at California Energy Company's Navy I pad boosted generating capability in the field to 75 megawatts by early 1989. Before the end of the year, completion of additional units at three other pads increased the capability to 230 megawatts, net, of electricity. (William Rintoul.)

oversee what is primarily The Geysers region in Lake, Mendocino, Napa and Sonoma Counties (in Santa Rosa).

In 1978, the Division took on more responsibilities when the State Legislature named it the lead agency under the California Environmental Quality Act for all geothermal exploratory projects on state and private lands in California. In addition, the Division was empowered to delegate lead agency responsibilities to counties that had geothermal elements in their general plans.

Throughout the years that followed, the geothermal industry grew into an important part of California's energy mix. In 1990, the installed capacity for generating plants represented about 2,480 megawatts, or enough electricity for a city of 2.5 million people. In equivalent terms, the geothermal resource was supplanting about 34.9 million barrels of oil per year, or about 95,400 barrels per day. In 1988, geothermal energy equaled about 3 percent of the total amount of energy used in California.

With installed capability of some 1,900 megawatts in 1990, The Geysers geothermal field is the world's largest geothermal development. California's other centers for the generation of electricity from geothermal resources in 1990 are the Coso geothermal field, which has an installed capacity of 230 megawatts; and the Imperial Valley's East Mesa, Heber and Salton Sea geothermal fields with an installed capacity of about 342 megawatts. The Casa Diablo field near Mammoth Lakes in Mono County is the site of a 7-megawatt facility and, to the north in Lassen County, 2.75 megawatts are generated from two geothermal power plants.

California's geothermal resources also have been tapped by shallow wells drilled to supply hot water for a variety of heating and recreational purposes, including greenhouses for growing roses, a municipal swimming pool, a church, public buildings and a correctional center in Susanville and environs in Lassen County; homes, an office and a gymnasium at the Fort Bidwell Indian Reservation in Modoc County; two schools and a hospital in Cedarville in Modoc County; an elementary school in Big Bend in Shasta County; the Indian Valley Hospital at Greenville in Plumas County; a public building in the City of Lake Elsinore in Riverside County; a plant nursery complex and ponds for fish and prawn farming near the Salton Sea in Imperial County; many spas in and around Desert Hot Springs, Riverside County; and the city hall and other public and private buildings in a large direct-heating project in the City of San Bernardino, San Bernardino County.

In an effort to preserve the Coso geothermal field's desert environment, development was restricted to certain areas. Joshua trees removed from plant and well sites were transplanted nearby and affected areas were reseeded with native vegetation, as shown on this hillside. (William Rintoul.)

11 The Environmental Age

The development of the oil industry in California was not unlike that of other industries. As the public became increasingly aware of the environmental impacts associated with oil, gas and geothermal development, the industry responded by making major infrastructural changes to address environmental concerns and comply with the California Environmental Quality Act. During this transitional period, the Division also placed increased emphasis on environmental concerns.

The habitat of this young San Joaquin Kit Fox, a California Endangered Species, was carefully preserved in the Buena Vista oil field during surface-cleanup operations, a special Division of Oil and Gas program. (Chevron USA Inc.)

While developers drilled wells at The Geysers to take steam out of the ground, they also put something back into the rugged countryside where they worked. They planted trees and native seeds to keep the environment as close to its original condition as possible.

The Union Oil Company of California activity of the early 1970s that found employees busily engaged in horticultural endeavors was a part of the growing realization, heightened by the rapid growth of California's population, of the importance of being good neighbors.

At one time, there had been some thought the stark, white insulation wrapping on steam lines was pretty, laying a neat spider's web over the geothermal field. The feeling was not shared by all. In deference to those who did not think the pattern of pipelines an improvement to the landscape, the company repainted existing lines the color of bayberry, which blended in with the natural foliage and soil.

Roads, drill sites and power plants were carefully planned and constructed to blend with the natural environment.

Replanting native grasses, trees and shrubs became major projects. For immediate ground cover to prevent erosion, workers planted wild oats, rye and other grasses, using the hydro-mulching system in which they seeded, fertilized, mulched and watered in one operation.

For long-term growth, such native plants were planted as bigleaf maple, chamisa, California buckeye, coastal whitethorn, western redbud, mountain-mahogany, toyon, chaparral pea, digger pine, coffeeberry, willow, California bay, California wild grape and several varieties of oak.

Some plants were spot seeded, fertilized and caged to keep deer from eating them. Other plants were grown as greenhouse seedlings for later transplant. Since many of the native plant seeds had built-in dormancy periods to prevent frost kills, they were refrigerated to bring them through the dormancy period and hasten the planting dates.

Actually, the environmental awareness campaign went back at least to war-time days of 1943 and an urban setting hardly more than three miles west of the old Los Angeles City field where wildcatters around the turn of the century had dug sumps in backyards and run oil down the streets without thought of consequences.

In the second February 1943 issue of *California Oil World*, the magazine reported, "Shell Pioneers Noiseless Well." The article said of the company's

In the 1920s, oil production with no hint of soundproofing or site landscaping shared the road with automobiles in urban Los Angeles. (Los Angeles Public Library.)

Verne Community No. 1 in Los Angeles, "The derrick and all machinery are swaddled like babies in fluffy rock wool, then enclosed in transite mineral board."

The exploratory well was being drilled on what was known as Gilmore Island at the corner of First and Gardner Streets. The magazine said the operator was making a contribution to the country by demonstrating that drilling could be carried on in populated areas. The well was a deeper pool test for the old Salt Lake field, a 1902 discovery. It had been named "Verne" in honor of Jules Verne, author of *20,000 Leagues Under the Sea,* "whose dreams once seemed fantastic but have been outdone by modern ingenuity." Shell was using "...electric drilling equipment housed in a fume-noise-odor proof covering and further isolated from sight by a tight high fence around the location."

Soundproofed derrick helped muffle the noise from drilling operations at Sansinena in the 1950s. (Unocal Corporation.)

To make oil development at Sansinena more acceptable, Union Oil Company of California put wellheads and other producing facilities in concrete-lined cellars, hiding them from view. (Unocal Corporation.)

Use of this shaft-driven rotary table reduced noise at the Gene Reid Drilling Inc. rig, drilling in the Sansinena field. A chain-driven rotary table was more commonly chosen. (William Rintoul.)

Though the Shell wildcat proved to be a dry hole at 7,924 feet, the pioneering effort at soundproofed drilling planted the seed of rig-site beautification and the idea of shielding operations from a sensitive public alienated by the legacy of oil-stained derricks, tanks and pipelines that had sprouted at Signal Hill, Santa Fe Springs and Huntington Beach during the booming 1920s.

In 1949, Union Oil Company secured a variance from the Los Angeles County Planning Commission permitting the company to drill directional holes from a small drilling "island" to develop the Sansinena field, a 1945 discovery near Whittier. The field lay in an aesthetically pleasing panorama of comfortable estate-like homes set in thriving avocado and citrus groves.

One of the contractors in the drilling that followed at Sansinena was Gene Reid Drilling Inc. When the Bakersfield contractor moved in to drill, rig-up time took six days instead of the normal 16 hours. Rig builders spent four of the six days covering the derrick and rotary equipment with soundproofing material consisting of two layers of vinyl-coated glass cloth with one-inch sheet fiberglass filling, especially heat-processed and quilted. Laps of the fire-resistant, washable soundproofing panels were securely fastened with three-inch safety pins. As a further safety measure, the laps were also wired.

Colored green on the outside to blend with the landscape, the soundproofing was bright yellow inside, bringing around-the-clock daylight operating convenience and safety for the drilling crew. On the outside, there were alternating orange and white strips at the top of the derrick as a warning to low-flying aircraft.

Inside exhausts were connected with two outside master mufflers through flexed tubing. The master mufflers were built especially to avoid any back pressure. The operation was set up so no fluid would be run onto the ground, and there was no graded sump as there would have been in a San Joaquin Valley oil field. Instead, there was a square, 80-barrel tank to serve as a repository for shaker diggings. There were two mud tanks, one for clean mud, another for waste mud. Vacuum trucks hauled away all waste.

Before preparations were completed to spud in the well, the approximate cost to the operator was $30,000, compared with about $5,000 for the preparations usually undertaken in other California oil fields.

As Union proceeded with development at Sansinena, the field became a show place for demonstrating that oilfield operations could go hand in hand with relaxed living. Once Cy Rubel, Union Oil's director of exploration, drove a group of clubwomen through the field. The women had protested that oil fields ruined the landscape. Rubel offered to buy a new hat for any of the women who could point out an oil well. He finished the tour without having to buy a single hat.

In Los Angeles, the city council in 1950 set the stage for an urban oil search by adopting a comprehensive zoning plan that permitted the development of oil fields in residential and business districts, but only from approved drill sites, with operators committed to soundproofed rigs, electric power and daytime-only delivery of materials. Waste material was

The adoption of a comprehensive zoning plan in 1950 by the Los Angeles City Council set the stage for the appearance of new landmarks in the form of soundproofed drilling rigs in the city's changing skyline. This Occidental Petroleum Corporation drill site in Sawtelle oil field is on the hospital grounds of the Veterans Administration Center in West Los Angeles. (Division of Oil and Gas.)

to be hauled to an approved disposal site.

As wildcatters fanned out in a sophisticated revival of earlier boom times, the art of camouflaging drilling and production islands rose to new heights. Operators worked to harmonize drilling and production equipment with existing structures. Well sites were disguised to look like everything from high-rise office buildings to a lighthouse.

Early in the rush to find more oil, one company built two new "sets" on a movie studio's lot. Unlike other sets that formed the backdrop for the filming of movies, the new sets were the real thing. They were drill sites landscaped to look like the well-kept grounds of an estate rather than components of a major oil field.

The Twentieth Century Fox studio lot was in the Beverly Hills field, an aging field discovered in 1900. The field had reached its peak in 1912 with a modest production of 680 barrels per day and then declined until by the early 1950s there were only two wells pumping fewer than 100 barrels daily. Through the first half century of its life, the field produced only slightly more than four million barrels from the Wolfskill zone at an average depth of 2,500 feet. Universal Consolidated Oil Company decided the script needed another ending. The company leased the movie studio's lot, erected a soundproofed rig and drilled an 8,562-foot well which came in flowing 525 barrels per day of 24-gravity oil from a Miocene pay almost 5,000 feet deeper than the original producing sand.

To develop the deeper production, Universal excavated two "islands" on the studio lot, surrounding the islands with reinforced concrete retaining walls varying in height from four to twenty-one feet. After drilling fifty-two wells, the company removed the derricks and screened the islands with plants and shrubbery, leaving no visible evidence of the production operation from adjoining properties.

From Beverly Hills the search spread to nearby Cheviot Hills, where Signal Oil & Gas Company's geologists identified a potential oil field in an area that boasted some of the Los Angeles area's most expensive homes, including those owned by John Wayne, Fred MacMurray, Nelson Eddy, Jeanette MacDonald and King Vidor.

Jim Wootan, the company's exploration chief, settled on a brushy ravine inside the Hillcrest Country Club's golf course as the ideal drill site. The exclusive country club was the province of an affluent

On the Twentieth Century Fox studio lot, Universal Consolidated Oil Company discovered deeper pool production that turned the declining Beverly Hills field into a major oil producer. Sun Drilling Company drilled the discovery well for Universal. (William Rintoul.)

Inside Sun Drilling Company's soundproofed rig on the Twentieth Century Fox studio lot, a crew set 7-inch casing to complete another Universal Consolidated Oil Company well. (William Rintoul.)

membership drawn from the worlds of finance, the professions and show business. What incentive could Signal offer to persuade the club to allow the company to drill for oil only a hundred yards from the clubhouse?

Jack Benny, one of the country club's members, quipped, "Perhaps if we sign up with Signal, we will be as rich as Bob Hope or Bing Crosby someday."

However, Signal found the Hillcrest directors concerned with rising maintenance costs, insurance, taxes and other overhead, and succeeded in getting permission to drill.

Signal also secured a lease on the nearby Rancho Park golf course, which was municipally owned.

To ensure that drilling from islands on the two golf courses would be acceptable, Signal's acoustical engineers enlisted the help of Hollywood's soundstage experts to soundproof the drilling derricks. Architect Henry C. Burge of the University of Southern California achieved the effect of making the derricks "inconspicuous, if not invisible" through the use of landscaping and color. To conceal the derricks from golfers, Signal under Burge's direction planted 60-foot-tall palms, plus Canary Island pines. The company enhanced the camouflage by painting the jackets grass-green at ground level, grading to sky-blue at the top.

If there were any need for proof that a 9,000-foot well could be drilled without noise, odor or detectable vibration, it came during the playing of the 1958 Los Angeles Open Golf Tournament at Rancho Park. Taut-nerved golf pros played critical shots in the shadow of Signal's drilling rig without being disturbed by the fact, if they were aware of it, that an oil well was being drilled just over the fence from the green.

Signal drilled fifteen wells from the Rancho site, thirty-three from the island on Hillcrest.

This drilling island in Boyle Heights oil field, a residential area, was a far cry from the forest of wooden derricks that competed with frame houses for space in turn-of-the-century Los Angeles. The wells at the site were later plugged and abandoned and the site was restored to its natural condition. (Richfield Oil Corporation.)

In the Las Cienegas area east of Beverly Hills, Signal teamed with Union Oil Company to lease oil and gas rights from nearly 25,000 home and property owners. Union, as operator, found oil. For initial development of the field, the company chose the parking lot of a shopping center on Pico Boulevard near the intersection with San Vicente Boulevard. After drilling was completed, Union built a small, office-type building over the production equipment, enclosed the rest of the site with a brick wall and landscaped the property. The field earned a degree of fame as the only oil field with an address: West Pico Boulevard. The site was one of five from which 122 wells were drilled to develop Las Cienegas field.

Another company joined the Beverly Hills play. Occidental Petroleum Corporation found oil one and one-quarter miles east of where Universal had proved up deeper pool production. The company enlisted a

Sheathed in soundproofing, Union Oil Company of California's Union-Signal-Pacific Electric No. 1 opened Las Cienegas field in Los Angeles. (Unocal Corporation.)

Hidden behind a decorative block wall at Adams Boulevard and Gramercy Place, sixteen Unocal wells unobtrusively produce oil from the Las Cienegas field underlying Los Angeles. (Unocal Corporation.)

team of architects to design its "oil derrick" on an urban drill site at the corner of a pair of major streets whose names evoked the memory of a pair of men who played roles in the early history of oil in California. The site was at the corner of Pico Boulevard and Doheny Drive. Andreas Pico in the mid-19th century had obtained oil from seeps found in a canyon near Newhall and distilled it for use as an illuminant at the San Fernando Mission. Edward L. Doheny had launched the Los Angeles City field boom before the turn of the century.

Occidental concealed what it termed "the world's first architecturally designed oil derrick" within what appeared to be a modern ten-story office building. The company enclosed the drill site with a 12-foot-high flagstone wall and landscaped the whole with trees, shrubs and ground cover. At a ribbon-cutting ceremony of January 20, 1966, Los Angeles Mayor Sam Yorty praised Occidental for the company's "outstanding contribution to civic beauty in a heavily-populated area of the city." Among those invited to attend the ceremony were the approximately 3,000 residents of the area who as lessors would share in royalty payments.

At a luncheon one month later in the Los Angeles Chamber of Commerce building, Robert S. Bell, chairman of the board of Packard Bell Electronics Corporation and an official of Los Angeles Beautiful, a nonprofit civic organization, hailed Occidental for its efforts in "going well beyond the call of duty in spending approximately $1,000,000 to build the novel sky-blue derrick."

The official statement said: "Los Angeles Beautiful salutes this new approach by Occidental Petroleum Corporation to instill beauty of design into every change in our environment, including oil exploration and recovery."

The company subsequently drilled 58 wells from the site.

Another discovery by Standard Oil Company of California proved up production one and one-half miles southeast of Occidental's "high rise."

On Pico Boulevard at Genesee Avenue in West Los Angeles, the company built the Packard drilling

Among those who greeted Dr. Armand Hammer, Occidental's chief executive, at dedication ceremonies in January 1966 for Occidental's "novel sky-blue derrick" in the Beverly Hills field were many of the approximately 3,000 residents of the area who as lessors would share in royalty payments. (Occidental Petroleum.)

A high wall shielded the Beverly Hills operation from passers-by. (William Rintoul.)

Standard Oil Company of California's Packard drilling structure housed the company's oil wells in West Los Angeles. (Chevron USA Inc.)

structure. Designed to look like a modern office building, the structure met design criteria so convincingly, according to one account, that two vice presidents of a rival oil company actually drove by the building three times before a police officer convinced them it was really Standard's "oil well."

The exterior of the 13-story building was finished in horizontal steel panels colored beige. Grounds were landscaped. Inside, the soundproofed structure housed two drilling rigs. The entire drilling operation was conducted indoors, including loading and unloading trucks, cementing and logging. An innovation was a viewing balcony from which residents of the neighborhood and visitors could watch the work going on inside. Standard drilled 80 wells inside the structure.

Before the decade of the 1960s ended, wildcatters had proved up 200

Standard's Packard drilling structure accommodated two drilling rigs, trucks, mud logger's trailer, pipe storage and other necessary production gear. (Chevron USA Inc.)

million barrels of new reserves beneath an area lying in a great arc from downtown Los Angeles to the waters of Santa Monica Bay. They discovered nine oil fields and found significant amounts of oil under three existing fields.

One of the new fields was Venice Beach, a 1966 discovery by Mobil Oil Corporation. The company drilled and completed six wells to develop the offshore lease from a drilling island on the ocean front. Mobil camouflaged the drilling structure to look like a lighthouse.

While Mobil's structure lighted the way for oil development at Venice Beach, the offshore search was spreading farther out to sea off California.

The discovery of the Summerland Offshore field in 1958, the first offshore field to be found in California as the direct result of offshore drilling, was followed by other discoveries on state tidelands, including Cuarta Offshore, The Texas Company, 1959; Gaviota Offshore Gas, Standard Oil Company of California, 1960; Naples Offshore Gas, Phillips Petroleum Company, 1960; Coal Oil Point Offshore, Richfield Oil Corporation, 1961; Conception Offshore, Phillips, 1961; Alegria Offshore, Richfield, 1962; Caliente Offshore Gas, Standard, 1962; Molino Offshore Gas, Shell, 1962; South Elwood Offshore, Richfield, 1965; and Carpinteria Offshore, Standard, 1966.

The development of the latter field brought the Federal Government into the picture. When it became apparent federally owned land beyond the three-mile limit would be drained by operations on an adjoining state lease, the Federal Government held its first lease sale off California, offering one drainage tract in December 1966.

Who said oil can't get along with sophisticated neighbors? Near the intersection of Olympic Boulevard and the Avenue of the Stars in the affluent sector of Beverly Hills known as Century City, a soundproofed rig in the early 1980s quietly rubbed shoulders with high-rise condominiums and high-rent apartments, the Century City Hospital and Medical Building, Beverly Hills High School and, dominating the whole scene, the Century City twin towers, rising 44 stories into the sky. (William Rintoul.)

A drilling rig, disguised as a lighthouse, was used by Mobil Oil Corporation to develop the Venice Beach Offshore field. (Division of Oil and Gas.)

A THUMS island. The covered drilling derricks are designed to look like high-rise buildings, and are moved on tracks circling the island. The island is landscaped with sculptured forms, some providing a frame for waterfalls that plunge to the ocean, as well as with shrubs and palm trees. (Division of Oil and Gas.)

Fourteen months later, the Federal Government held another sale, offering 110 tracts in opening the Outer Continental Shelf (OCS) in the Santa Barbara Channel to oil exploration. In the final bid of the sale, a bidding group composed of Gulf Oil Corporation, Texaco Inc., Mobil Oil Corporation and Union Oil Company of California won Tract 402 with a bonus of $61,418,000, or $11,373.70 per acre, which was the highest of the sale and the largest single bid ever offered for an offshore lease.

The wells drilled in federal OCS waters were regulated by the Federal Government, rather than by the Division of Oil and Gas. One year later, the fifth well to be drilled in federal waters from Platform A, which was positioned to develop what was later named Dos Cuadras Offshore field, blew out while the crew was preparing to log at 3,203 feet. The crew accidentally dropped the string of drill pipe down the hole and closed the blind rams to shut in the well. Soon after, oil and gas began to bubble to the ocean surface from a fracture line on the ocean floor. Ten days after the flow of oil and gas began, the operator succeeded in controlling the blowout by pumping 13,000 barrels of 90- to 110-pound mud down the hole. The coup de grace was administered with 900 sacks of cement, plugging the well for abandonment.

In its wake, the oil spill left a legacy of cancelled lease sales, lawsuits and proposed legislation to block offshore drilling in both state and federal offshore areas.

Although over 1.5 million feet of hole for some 925 wells had been cut since 1955 in state offshore areas under Division of Oil and Gas regulation from offshore platforms and islands on the state's tidelands leases without any serious problems, the Division chose not to be complacent. Responding to the spill on federal land, a special Division Offshore Unit was formed, effective July 1, 1969. The Offshore Unit assumed the supervision of all offshore drilling and production operations in waters within state jurisdiction.

Division personnel began aerial surveillance of offshore operations and established liaison with other state and federal agencies to ensure early detection of oil leaks and to note natural offshore seep activity.

An early assignment was the preparation of a map showing all natural oil- and gas-seeps in the offshore areas of California. The study of seeps led to the publication in 1971 of a paper titled *California Offshore Oil and Gas Seeps*. The paper offered the first comprehensive report on the state's offshore seeps, and documented 54 such occurrences.

In May 1974, new offshore administrative regulations prepared by the Offshore Unit became effective. The regulations included requirements and procedures for all phases of California's offshore oil- and gas-well operations. The regulations also allowed for the establishment of field rules for areas

Shell Oil Company received the Division of Oil and Gas' "Outstanding Oilfield Lease and Facility Maintenance Award" for its Fulton and RHD leases and the D & E Sand Unit Steam-Generation Facility in the Midway-Sunset field. (Michael Glinzak.)

with known geological and engineering conditions determined by previous drilling.

Concurrent with the Offshore Unit's surveillance of offshore activities, the Division focused onshore environmental enhancement efforts on identifying and eliminating oilfield sumps that could be hazardous to wildlife or usable groundwater supplies. The program called for the Division to coordinate field inspections and aerial surveys with the efforts of other key state agencies and private groups such as the Audubon Society.

In 1971, the Division of Oil and Gas and the California Department of Fish and Game began an inventory of oilfield sumps. The two agencies identified 4,065 sumps as hazardous or immediately dangerous to wildlife. Although some of the sumps were screened or eliminated on a voluntary basis, on January 1, 1974, Assembly Bill 2209 became effective. Under its provisions, a legally mandated, full-scale sump inspection and correction program began. The Division was empowered to serve legal notice to the operator or owner of a hazardous sump, as identified by the Department of Fish and Game, and to take remedial action with a deadline for correction. If corrective action was not taken within a reasonable time, the State Oil and Gas Supervisor could order the closure of the producing operation maintaining the sump.

In a joint report to the State Legislature in March 1979, the Division and the Department of Fish and Game stated that all but 88 hazardous sumps had been screened or eliminated in California. The report said action would be taken to correct the remaining sumps as expeditiously as possible, and that they expected a small number of sumps would be discovered each year, either new sumps or old ones where screening had deteriorated. The goal of the Division of Oil and Gas and the Department of Fish and Game would be to keep this number as low as possible, the report pledged.

The matter of sumps was not the only item on the Division's environmental agenda. In 1975, the Division targeted hazardous and idle-deserted wells for abandonment under a program to be financed in part by Division funds, and in part by drilling bonds. The assignments in the years that followed led Division personnel back into the past to close the book on chapters from some of the oil industry's earliest operations.

In the San Joaquin Valley, nine gas wells that had been drilled beside the Tuolumne River in the years between 1926 and 1946 were flowing large volumes of saltwater into the river, aggravating already critical salinity conditions. A study by the Central Valley Regional Water Quality Control Board concluded the wells were pouring an estimated one-half million pounds of salt per year into the river, causing about $3 to $5 million per year in damage to agriculture. The Division's District 5 office supervised site preparation, logging and well abandonment that

Soundproofed drilling made its debut in the San Joaquin Valley in 1953 when Texaco moved into the rural town of Arvin to look for an extension of the Mountain View field. (William Rintoul.)

stopped the flow of salt water. The cost for the operation was almost $140,000, of which about $31,000 was contributed by the State Water Resources Control Board, the remainder by the Division.

In the Coastal Region, a well in the Sespe field had been leaking oil into Sespe Creek in the California Condor Reserve for many years. The well was one of twelve drilled in the remote Green Cabins area between 1901 and 1937. Production from the area had ended in the 1950s. The abandonment operation involved the combined efforts of the Division's District 2 personnel and the U.S. Forest Service with the assistance of several oil and gas service companies that contributed funds and services. Because of the rugged terrain, it was necessary to transport the rig, equipment, supplies and personnel in and out of the site with a Sikorski Sky Crane helicopter. Total cost of the project was $138,000.

In the old Summerland field in Santa Barbara County, District 2 personnel designed an abandonment program for 26 wells and supervised the abandonment operations. Most of the wells had been drilled before 1894 and had been left around the turn of the century with only a covering of dirt. Working with county officials, Division engineers located the wells two- to-nine feet below the surface on six ocean lots. Because of power lines, it was necessary to use a crane for some of the work.

In the Los Angeles Basin, the City of Brea uncovered an open well only a few feet from State Highway 142 and up-slope from the Carbon Canyon stream bed. The investigation by the Division's District 1 staff revealed the well was full of oil and leaking a small amount of gas. The Division identified the well as one that was believed to have been abandoned in early 1915. When reabandonment work began, the well proved difficult to reenter. A large auger rig was used to clean out the first 80 feet, recovering casing protectors and collars, valves, timber, cable and a cable-tool drilling bit. Two drilling bits and four mills were worn out before a depth of 1,064 feet was reached, and the hole was reabandoned by filling it with cement.

In the Huntington Beach field, Southern California Gas Company during routine transmission line inspections in the city found gas leaking from a well that had been drilled in 1947 and abandoned in 1953. Reabandonment operations under the Division's program went smoothly until the well was to be plugged with cement. Then tests showed that methane gas was leaking from the space between the casing and the well bore. Examination of the well's electric log by District 1 personnel indicated the gas was coming from a shallow zone at about 83 feet. Tests showed the gas probably was biogenic in origin, that is, formed by bacterial action, and was not from the oil and gas reservoir penetrated by the well. On the Division's recommendation, the City of Huntington Beach decided to use the well to vent the biogenic gas. To do so, the well was perforated at 83 feet, and a hollow light standard was attached to the well casing. The gas was vented through the well and light standard to a height of about

This screened oilfield sump, properly maintained by Chevron USA Inc., is on the McNee lease in the Buena Vista oil field. An old, wooden derrick stands in the distance. (David Clark.)

40 feet above street level. Since the well was located in the front yard of a house in a residential area, black plastic sheeting was laid out during the abandonment operations to protect the street, driveway and lawn.

Along with concern for the elimination of hazardous sumps and wells, the Division's environmental program boasted other facets. One was joint sponsorship with the Department of Fish and Game, Western Oil and Gas Association, Audubon Society and Western Interstate Commission for Higher Education of a research project to provide oilfield operators with practical information necessary to preserve and enhance wildlife habitat in the southern San Joaquin Valley. The Division published the study's findings to assist operators in recognizing wildlife habitat needs and methods of establishing wildlife cover, food and water in the proper interrelationship. In another project, the Division developed an environmental lecture program that was presented to various groups in the petroleum industry, governmental agencies, schools and such private organizations as the Audubon Society and Environmental Defense Fund.

Emphasis on the environment led to another Division program in the 1980s. It began on March 24, 1982, when three oilfield operators in District 1 received the first Division awards for "Outstanding Oilfield Lease and Facility Maintenance." The awards were presented by M. G. Mefferd, State Oil and Gas Supervisor, to Conoco Inc. for leases in the Seal Beach field, Getty Oil Company for the Vickers lease in the Inglewood field and George A. Jones for his lease in the East Coyote field.

The awards recognized what Mefferd described as "...those operators who have gone the extra mile to be good neighbors." He emphasized the awards were not given just for complying with regulations. "All operators are expected to do that," Mefferd said. "These awards are given to those operators who make the extra effort and go to the extra expense to make their leases a showcase. They are truly fine representatives of the oil industry."

Every year since, more oil, gas and geothermal operators have joined the award-winning ranks for maintaining their leases in an exceptional manner.

The Division's construction-site review and well-reabandonment program was begun in December 1986. The program prevents construction over improperly abandoned wells, most of which are old wells that were abandoned before the Division was founded. Program procedures were sent to cities, counties and special agencies that issue building permits. Under the program, hundreds of construction site plans have been reviewed and many wells reabandoned.

A program to remove unneeded equipment, such as cables, flowlines, pipelines, and tanks, from oil fields on the western side of Kern County was begun in 1989 under the auspices of the Division of Oil and

The oil field's ubiquitous pumping unit often was referred to by many field hands as a grasshopper pump. In the Santa Maria field, Unocal rigged up a unit on the Bradley property beside Highway 101 to fit the image. (William Rintoul.)

Gas. Buena Vista oil field, discovered in 1909, was the first to be cleaned up under the program.

The effort at Buena Vista was especially significant because the 50-square-mile oil field contains large areas of prime habitat for several listed endangered species, including the San Joaquin Kit Fox, the Giant Kangaroo Rat, and the Blunt-Nosed Leopard Lizard. The greatest logistical problems in the cleanup were created by the presence of these endangered species in the work areas. Consultations and field inspections were conducted with the U. S. Fish and Wildlife Service and company biologists to develop work plans that protected the environment during cleanup activities. As a result of the program, oil companies operating in Buena Vista field spent about $1.5 million to remove 4,900 tons of unneeded material.

During the Buena Vista field surface-cleanup activities, habitats of several endangered species were carefully preserved. (David Clark.)

Some of the 4,900 tons of unneeded material removed from the Buena Vista oil field. (David Clark.)

Buena Vista oil field, after cleanup activities. (David Clark.)

12 At Work

In Woodland on June 11, 1990, Division of Oil and Gas personnel posed for a family portrait similar to that taken 73 years and 6 months before. Standing in the back row, left to right, are Richard K. Baker, district deputy, Long Beach; Richard F. Curtin, district deputy, Coalinga; Richard P. Thomas, geothermal officer; Patrick J. Kinnear, district deputy, Ventura; and Robert A. Reid, district deputy, Woodland. Standing in the middle row, left to right, are William F. Guerard, Jr., technical services manager, Sacramento; and Hal P. Bopp, district deputy, Santa Maria. Seated, left to right, are David C. Mitchell, senior oil and gas engineer, Bakersfield; M. G. (Marty) Mefferd, state oil and gas supervisor, Sacramento; and Kenneth P. Henderson, chief deputy, Sacramento. (Division of Oil and Gas.)

One recent morning in the San Joaquin Valley, Kathy Roush left the Bakersfield office of the Division of Oil and Gas to drive to the Yowlumne oil field, 20 miles to the southwest. The purpose of the trip was to inspect the blowout prevention equipment at a contract drilling rig on location to drill a 12,500-foot development well.

The drive to Yowlumne took Roush, an Energy and Mineral Resources Engineer with the Division, out Old River Road past fields of cotton and alfalfa and, farther along, packing sheds clustered by the railroad spur siding known as Conner Station, a shipping point for agricultural produce. There she passed within a few miles of the Paloma oil field, a 1934 discovery and the site of a deeper pool exploratory well that 36 years before had riveted the attention of the nation's oil industry. For almost three years, the 21,482-foot hole had been the deepest in the world. Unfortunately, the drill bit had failed to find commercial production, and the Paloma deep test had been consigned to service as a water-disposal well, clinging to a shred of fame as the last well in California to hold the world's depth record.

Six miles past Conner Station, the paved highway took Roush through the Rio Viejo field, where the search for deep production had proved more fruitful than at Paloma. At Rio Viejo, a discovery two decades after the Paloma failure, Christmas tree wellheads marked the ten wells through which high gravity oil and natural gas flowed from the Stevens sand some 14,100 feet below the surface, qualifying the field as the site of the deepest commercial production in California.

Just short of the California Aqueduct overpass about one mile down the road, Roush made a right turn to follow a hard-surfaced secondary road out into farmlands providing the setting for the Yowlumne field. The field, a 1974 discovery, had taken its name from a suggestion by the late Richard C. Bailey, director of the Kern County Museum, to name the field after an Indian tribe he described in *Heritage of Kern*, a book published in 1957 by the Kern County Historical Society. The name, meaning "wolf people," was that by which the first inhabitants of what now is Bakersfield had been known among their fellow Yokuts.

The mast of the powerful drilling rig toward which Roush drove towered above a cotton field, visible for miles as the only rig at work in the almost fully developed oil field. Pulling up at the graded site, Roush donned her hard-hat and, clipboard in hand, checked in with the drilling specialist serving as the operator's representative at the well site. After a brief discussion, she made her way past mud pumps and rig engines to the steel substructure that elevated the rig floor some twenty-five feet above the ground. In the space between the rig floor and the concrete well cellar, the blowout prevention stack had been installed as the first conduit through which the drill pipe would pass in the drilling of the well. The stack consisted of a series of four large, heavy-duty devices, one above the other, that would provide the capability to shut off the escape of gas or other fluids if the well should attempt to blow out. In recognition of the pressures that might be encountered by the well, the stack was designed to hold back a flow

Kathy Roush met with Bill Works, the operator's representative, at the drill site where a contractor's rig, background, was drilling the Yowlumne well. (William Rintoul.)

coming at it with pressures as high as 5,000 pounds per square inch.

To the inspection task, Roush brought solid credentials. A graduate of California State University, Northridge, with bachelor's and master's degrees in geology in 1983 and 1986, respectively, she had worked between times in school for a consulting engineering firm in Bakersfield. Her interest in geology and engineering that led to involvement in the oil industry stemmed from a high school instructor's lecture on plate tectonics when she was a junior at Cleveland High School in Northridge, California. Since joining the Division of Oil and Gas in 1986, Roush's assignments had included, in addition to blowout prevention equipment tests and inspections, the witnessing of well abandonments to ensure they conformed to state standards, environmental inspections, record keeping and, when time was available, the preparation of technical papers for Division publication. She was presently working on a pair of papers, one on the depositional environment of the Domengine Formation in the Coalinga area and another on the Antelope Shale in the Midway-Sunset field.

The blowout prevention stack was bolted to a string of 10-3/4 inch surface casing cemented to a depth of 1,505 feet in the Yowlumne well. The stack included three ram-type preventers and an annular preventer. The bottom preventer on the stack was a pipe ram, designed to fit around the drill pipe so the preventer could be closed when drill pipe was in the hole. The preventer installed above the pipe ram was a blind ram, designed for full closure once the drill pipe was out of the hole. On top of the blind ram was another pipe ram. The annular preventer, also known as the bag, was installed above the three ram-type preventers. It could be closed around any object or fully closed when no drill pipe was in the hole. Each preventer was operated hydraulically with the touch of a switch, making it possible to close the proper preventers in a matter of seconds. One set of controls was located next to the driller on the floor of the drilling rig. A duplicate set was positioned by the accumulator, which by regulation had to be at least 50 feet from the wellbore. Between the lower pipe ram and the blind ram there was a spool, a device with two

The blowout prevention equipment that Kathy Roush was assigned to inspect was designed to handle pressures as high as 5,000 pounds per square inch. (William Rintoul.)

The blowout prevention equipment was housed in the space between the concrete cellar and the rig floor, more than two stories above ground level. (William Rintoul.)

connecting lines. One was a kill line through which weighted mud could be circulated into the hole to kill a "kick," which occurs when formation fluids enter the well bore once formation pressures exceed those exerted by the weight of the drilling mud. The other connecting line was a choke line through which formation fluids could be circulated out of the hole through a series of valves.

The testing procedure called for the rig's crew to pressure-test the stack of blowout preventers to simulate conditions that might be expected if the well should attempt to blow out. By closing valves, each preventer as well as various valves on the kill and choke lines and the drill string could be tested for leaks. After the preventers and other equipment had been tested, the procedure called for a visual inspection of the accumulator that contained the hydraulic fluid used to open or close the stack and the emergency back-up system for the accumulator, powered by nitrogen in case the rig's electrical system failed.

About an hour after arriving at the drill site, Roush was ready to return to Bakersfield, assured that the blowout prevention equipment met all standards.

While the Yowlumne well was on the threshold of what its operator hoped would be a long and productive life, another Kern County well was reaching the end of its productive span.

Joe Perrick, another Energy and Mineral Resources Engineer with the Division's Bakersfield office, headed out to the Kern River field to witness one step in the abandonment of the well. Perrick, from Butte, Montana, had graduated from Carroll College in Helena, Montana, with a bachelor's degree in mathematics in 1982 and from Montana College of Mineral Science and Technology in Butte with a degree in petroleum engineering in 1984. He had gone to work as a field engineer for an oilfield service company in Williston, North Dakota, and after a stint in Tulsa had been assigned to the company's Bakersfield office in 1985. He had worked for another service company in Bakersfield as a reservoir engineer before joining the Division of Oil and Gas.

Ninety years before Perrick's morning run to Kern River, the new-found field, proved up by a hand-dug hole, had been separated from Bakersfield by an open belt of farm and grazing lands. One of the great sporting events of the field's early days had been a foot race in February 1901 between an oil worker with the Four Oil Company known as the Roadrunner and the Kentucky Kid with the West Shore Company, from the oil field to the Southern Hotel in Bakersfield. The Roadrunner had outdistanced the Kid by fifty yards, covering the seven miles in forty-six minutes. Hundreds of dollars in bets

At the Kern River well scheduled for abandonment, Joe Perrick, left, and Paul Owen, the well-servicing contractor's foreman, went over the program for the cavity shot. (William Rintoul.)

had changed hands.

In the years that followed, both the city and oil field had spread out until the two met at the foot of China Grade where some of Bakersfield's most luxurious homes dominated prestigious Panorama Drive. For Perrick, the trip to the Kern River field was largely on urban streets, culminating with a circuitous trip to the Mecca lease over oil-company roads past a goodly number of the field's more than 8,500 wells. A well-servicing rig marked the site of the well that Perrick sought, easing his task of finding the well in the maze of more than 50 wells on the 80-acre property.

The well that was to be abandoned had been completed forty-one years before by a small independent operator from a depth of 1,120 feet for 10 barrels per day of 12.8-gravity oil. The decision to abandon the well did not reflect a conclusion that all the oil had been recovered. In fact, after four decades of productive life, the well actually was producing more than twice as much oil as in the beginning, averaging 26 barrels per day.

The reason for the increased production and, paradoxically, for the abandonment was steam. Starting with pilot tests in the early 1960s, the injection of steam had revolutionized the field, increasing production from a pre-steam level of about 23,500 barrels per day to about 125,000 barrels per day as the decade of the 1980s drew to a close, making the field the fifth-most productive oil field in the United States.

The major company that had acquired the Mecca property from the independent owner in the early days of the turn to steam was in the process of expanding a steamflood project. When a check of the casing in the well in question raised doubts about the well's mechanical integrity, the operator decided to abandon the well and drill a replacement well.

Before the morning when Perrick arrived to witness the latest step in the abandonment procedure, several steps already had been taken. A well-servicing crew had bailed the well clean to the original depth of 1,120 feet and set a cement plug from the bottom of the hole back up to 1,030 feet, using a slurry that included 35 percent silica flour, 10 percent gypsum and 2 percent calcium chloride to ensure a plug that would resist thermal degradation.

After the plug had set, a wireline service specialist licensed to handle explosives had detonated a 4-inch by 10-foot, 50-pound explosive canister at the interval from 1,020 to 1,030 feet to create a cavity. Afterward, the rig crew had run a string of tubing into the well and set another cement plug back up to a depth of 860 feet, filling the cavity and isolating the zone that was going to be steamed.

The next step, which Perrick had come to witness, involved another "cavity shot," this one from 670 to 680 feet. At the well site, he talked with the well-servicing contractor's foreman on the job. After confirming that the top of the cement from the first cavity shot was at 860 feet, Perrick watched while the

Joe Perrick with Gary Dye, right, contract explosives expert. The 50-pound canister in the background is ready to be run into the well on a wireline. (William Rintoul.)

contract explosive expert ran a 50-pound explosive canister into the well on a wireline, carefully positioning the charge at the desired interval. The explosion could scarcely be felt on the surface as water above the canister in the wellbore provided a cushion, directing the energy from the explosion horizontally into the formation. The wireline was reeled in, bringing the canister that had carried the explosive to the surface. Only a few shreds of metal remained.

Perrick approved the cavity shot. The crew of the well-servicing rig began running tubing into the well while a service company's cementing truck stood by to pump in another cement plug, this one to be spotted up to a depth of 560 feet.

The next step would be to fill the well's casing with a sand-cement mix back to the surface. The casing then would be cut off at least five feet below the surface and a steel plate with the well's name beaded on it would be welded over the casing. The final step would be to clean up the surface location, restoring it to its natural state.

Depending on following days' assignments, Perrick might make as many as two more trips to the well site. One return trip would be to inspect the final plug to verify that the top of the plug was within five feet of the surface. The second trip would be to witness renovation of the well site. Then, pending completed paperwork, a letter would be sent to the operator with final approval of the abandonment, officially closing the book on the well.

On the same morning that Joe Perrick was witnessing the cavity shot at the Mecca well, Dwight Isenhower also was busy in the Kern River field.

Isenhower, an Oil and Gas Technician in the Bakersfield office, had come to the Division in July 1979 in Sacramento after four years with the State Employment Development Department. He had grown up in Folsom, California, graduating from Folsom High School in 1974. After a little more than one year in the Division's data processing section in Sacramento, he had transferred to Bakersfield in September 1980, and completed courses in geology and mathematics.

Following approval of the cavity shot, floor hands on the well servicing rig prepared to run tubing so a cement plug could be pumped into the well. (William Rintoul.)

Cementers, right, stood by to pump cement into the well that was being abandoned. (William Rintoul.)

The flow chart at the wellhead in the Kern River field indicated the operator was injecting approximately 35 barrels of water per minute. (William Rintoul.)

One of the first steps in Dwight Isenhower's inspection was to read the gauge that recorded surface pressure to make certain the pressure did not exceed the allowable limit, ensuring that the water did not fracture the zone into which it was being injected. (William Rintoul.)

As he made his inspection, Dwight Isenhower put a check in the appropriate space on the worksheet, ensuring that all aspects of the well and well site were evaluated. (William Rintoul.)

The assignment that brought Isenhower to the Kern River field, like that which accounted for Perrick's presence, related to the use of steam to increase recovery of heavy crude oil.

The positive aspect of steam's role had been the huge increase in oil production that promised the eventual recovery of as much as seventy-five percent of the original oil in place, instead of the ten to fifteen percent that might have been recovered without thermally enhanced oil-recovery projects.

A less positive side was the huge increase in the volume of water recovered along with oil from Kern River's wells. When the first steam injection projects began in the early 1960s, the field was producing about five and one-half barrels of water for each barrel of oil. As production of oil increased, water production moved ahead until by the end of the 1980s the volume of water amounted to roughly nine barrels for each barrel of oil recovered.

It had been one thing in the 1960s to dispose of the approximately 130,000 barrels per day of water produced along with oil. It was quite another to dispose of the 1.1 million barrels per day produced with the field's oil in the late 1980s. While some of the water could be cleaned up and used as feed stock for the generators that produced steam for injection, there was still a large volume left to be disposed of in an environmentally acceptable manner.

An obvious solution was to put the water back in the ground. The method had been employed as early as 1939 in Kern County's Greeley field. By 1964, when the first water-disposal well went into operation in the Kern River field, there were 127 such wells in 75 California fields disposing of some 260,000 barrels per day of produced water. From a single well serving as a conduit for the disposal of about 93,500 barrels per day of produced water during its first months of operation, the roster of water-disposal wells in the Kern River field had grown to 74 wells, returning about 265,500 barrels per day of water to the earth, or about one and one-half barrels out of each ten barrels of the 1.8 million barrels of produced water disposed of daily in the state's fields.

It was Isenhower's assignment, shared by Bill Winkler, an Energy and Mineral Resources Engineer in the Bakersfield office, to survey each of the approximately 600 water-disposal wells in District 4 once a year. The assignment also called for checking each of the district's approximately 1,800 waterflood wells once every two years and each of the district's approximately 4,800 steamflood wells once every five years, which meant about 2,500 inspections each year.

One of the first stops for Isenhower on this particular day was a major company's water-disposal well in the southern portion of the field. The flow chart at the wellhead indicated the operator was injecting about 35 barrels of water per minute. Isenhower read the gauge that recorded surface pressure to make certain the pressure did not exceed the allowable amount. The purpose of setting a limit was to make sure the zone into which water was being injected could receive the water without being fractured.

Isenhower also checked the gauge that measured pressure on the well's annulus between the tubing and casing. A show of pressure could indicate that water was migrating into the casing above the packer that had been set to confine water entry to the injection zone.

As he made his inspections, Isenhower put a check in the appropriate space on the worksheet on his clipboard. There was space for entries covering every aspect ranging from injection pressure to the cleanliness of the well site and whether there was an acceptable identifying sign on the well.

The worksheet included a place for evaluation of equipment in use, with space provided for a check opposite "good, fair or poor."

Before leaving the location, Isenhower carefully inspected the site for any sign of leakage. When he finished the day's inspections, he would enter the information in a computer, making it possible to retrieve the record when and if it might be needed. If a check appeared opposite a space such as "flowing leakage" or "no gauge," the operator of the well would be notified by a deficiency letter stating a given number of days within which the operator would be required to remedy the situation.

The Kern River field, of course, was only one of many California fields where oil producers had turned to disposal wells to solve the problem of returning unwanted water to the earth.

Another field with a water-disposal problem was Edison, a 1928 discovery three miles east of Bakersfield, where water production had climbed steadily. The Edison field had seen its peak production in 1953, with an average of 18,225 barrels per day of oil.

Then, along with each barrel of oil, the field produced almost two barrels of water. By the late 1980s, oil production had declined to 2,800 barrels per day, while water production had climbed to 33,600 barrels per day, or 12 barrels of water for each barrel of oil.

Concurrent with Dwight Isenhower's inspection of water-disposal wells at Kern River, Bill Winkler drew another assignment at Edison. The assignment stemmed from an independent producer's proposal to convert an existing production well in the Race Track Hill area of the field to serve as a water-disposal well.

The 5,421-foot well had been drilled in 1945 and completed to production flowing 1,400 barrels per day of 41.7-gravity oil and one million cubic feet per day of gas. The production had come from perforations in the interval from 4,687 to 4,692 feet in the Jewett sand. With the passage of time, the well's production had declined, leading to the owner's decision to recast the well as a water-disposal well.

In making the proposal to the Division of Oil and Gas, the producer had had to satisfy numerous conditions before arriving at the stage that called for Winkler's presence at the well site.

The producer had been required to submit extensive data on the area's geology and other pertinent information ensuring that the water-disposal project would not contaminate other zones.

The Division, after reviewing the data to its satisfaction, had placed a legal notice in the local newspaper, in this case, *The Bakersfield Californian*, inviting public comment on the proposed project. The notice, which was a requirement of the Division's Underground Injection Control Program, was followed by a 15-day period for public comment. In addition, the Division also had sent a letter to the Regional Water Quality Control Board (RWQCB), inviting that agency's comments.

After the waiting period, the Division had sent a draft approval letter to the RWQCB. Following its approval, the Division had approved the project. One requirement of the Division's permit was that within 90 days of the start of injection, a mechanical integrity test of the injection well would have to be made—and the results approved by the Division's representative.

Bill Winkler, left, and John Oldham, foreman for the independent operator that filed notice to convert an exhausted producing well to serve as a water-disposal well. (William Rintoul.)

The reason for Bill Winkler's trip to the Race Track Hill well was to witness the required mechanical integrity test. (William Rintoul.)

With the go-ahead from the Division, the producer had set a cement plug above the original producing interval in the well and shot injection perforations in the interval from 4,596 to 4,606 feet in the Pyramid Hill sand, which lay above the Jewett. An impervious layer of shale separated the two sands.

The reason for Winkler's trip to the Race Track Hill well site was to witness the survey of the injection well as required by the project permit. The purpose of the survey, known as a mechanical integrity test, was to make sure the injected fluid was going only into the zone for which it was intended.

Winkler had joined the Division after a six-month stint with a major oil company in Kern County's Cymric oil field. A graduate of Highland High School in Bakersfield in 1975, he had continued his education for two years at Bakersfield College before enrolling at California State University, Chico, from which he had graduated in 1981 with a degree in geology. At the Division, he had worked in both the Ventura and Bakersfield offices.

At the well site, Winkler found a contractor's wireline truck rigged to do the necessary survey work. The wireline had been run in the well. At the end of the line was a combination of five tools needed for the survey. The tools would be activated as desired by changing the polarity and the voltage on the connecting wire in the cable. A two-man crew was on hand, including the contractor's logging engineer and the contractor's operator. Also at the site was the production foreman for the operator.

The first step in the survey involved activating one of the tools on the wireline to make a continuous log of the well's temperature profile. The purpose of the temperature log was to detect intervals that may be invaded by injected fluids. Winkler witnessed the log run to confirm the mechanical integrity of the well.

The next step involved two other tools on the wireline. One was an injector through which small amounts (slugs) of radioactive iodine could be released into the fluid being injected. The other was a gamma-ray detector that would track the fluid movement by tracing the iodine marker. The radioactive iodine had an 8-day half life, which meant that in 16

Bill Winkler, left, and Stan Byrom, one of two members in the crew performing the mechanical integrity test, watched intently for a "spike." (William Rintoul.)

By passing the mechanical integrity test, the Race Track Hill well earned approval to continue service as a water-disposal well for another year. Left to right, Bill Winkler and Martin Lanzer, the contractor's logging engineer. (William Rintoul.)

days, only one-quarter would be left, and that would disappear quickly.

Before the tracer survey was run, the contractor's operator ran a background log, using the gamma-ray detector, from below the perforated injection interval to the top of the interval to get a gamma count, establishing a base line before beginning the next test.

Following completion of the background log run, the tool was run from the bottom upward at varying depths past the perforated injection interval to track the flow of injected fluid and monitor the points at which the interval accepted the fluid.

After this was checked, the operator moved the gamma-ray detector tool some 25 feet above the top perforations and released a slug. There was the expected "spike" on the recording chart as the radioactive slug passed the detector on the way down. The operator and Winkler carefully watched for about five minutes to see if there would be another spike, which would mean that fluid was coming back up outside the casing, indicating a failed test. There was no other spike, which indicated the fluid was going into the injection zone.

After the successful completion of the top perforation check, the operator moved the tools up the hole to position them in the tubing above the packer, which had been set at a depth of 4,458 feet. To test the packer, the operator then basically repeated the procedure used to test the top of the perforated injection interval, setting the tools some 25 feet above the packer, putting out a shot of iodine tracer, and waiting to see if it came back up past the packer.

The final check was of the integrity of the tubing in the well. The operator made a spinner count, using a fifth tool in the string run in the hole on the wireline. The tool had a propeller-like device to check the flow of fluid being injected at a rate of about 1,900 barrels per day. The count was 148 revolutions per minute (rpm) on top, 152 rpm on the bottom, indicating the tubing was mechanically sound. If the count on top had been drastically larger, it would have meant there was leakage, indicating there was a hole somewhere in the tubing.

About two hours after he had arrived at the well site, Winkler confirmed that the well had passed the mechanical integrity test. The independent producer could continue injecting fluid for one year before another survey would be required.

For Syndi Pompa, an indispensable tool for making rounds in District 4's oil fields was the large map book she carried with her in the four-wheel-drive vehicle.

The book contained a map for each of California's fields, showing the precise location of each well against a grid of section, township and range lines.

Before leaving the Bakersfield office on this particular morning, Pompa, an Oil and Gas Technician, consulted the map, lining out the best route to follow to arrive at the leases she meant to inspect. Her current assignment with the Division was to perform environmental lease inspections.

While the page in the map book showed the location of each well, it did not include—it could not include—the road that led to each well. That would have required an immensely large, immensely detailed map. It remained for Pompa to get on the scene and then, guided by a scattering of operators' signs and well

Numbers on the "horse's head" at the end of the walking beam helped Syndi Pompa find the wells she was looking for on her inspection. (William Rintoul.)

numbers, find the pertinent properties. She quickly found the fee property in the McVan area of the Poso Creek field, some 10 miles north of Bakersfield, that was her destination.

The Poso Creek field, a 1920 discovery that had produced 80 million barrels of oil, virtually all of it heavy crude, had fallen on hard times with the free-fall of crude prices in 1986. More than two thirds, or 660, of the field's wells were idle, with only 273 still pumping. Virtually all of the wells that dotted the hilly setting of the major oil company's lease that Pompa was assigned to inspect were idle, pumping units still, well sites empty of workers.

To the inspection job, Pompa brought the perspective of someone who only a few months before had been in the position of remedying environmental problems rather than finding them for others to correct. She had joined the Division after working three and one-half years for a major oil company in the San Ardo field, initially as a roustabout, later advancing to be a thermal project operator. Her chores had occasionally involved cleaning up well sites in compliance with Division requirements.

Now, her assignment with the Division was to pinpoint problems for others to address.

The silent lease she inspected, going from well to well, pumping unit to pumping unit, looked like a spic-and-span operation that might only the day before have been deserted by workmen after they had put everything in order. There were no oily rags in sight, no leaking oil around wellheads, no discarded work gloves or trash from lunch pails.

As she made the rounds, Pompa meticulously filled in the blanks opposite classifications on the environmental lease inspection worksheet, noting the operator's name, the field, the lease, the section, township and range locations and the well number or tank setting designation. There were questions to be answered. Did the well have a cellar? Was the cellar covered? Was the cellar full of fluid? Debris? Was the stuffing box leaking? Oil on the ground? Was there a well sign? Was there fencing around tank settings? Gathering installations? Was there a berm to hold any oil that might escape from a ruptured or leaking tank? If there were plugged and abandoned wells on the property, was restoration completed at the site? Was there a sump or sumps on the property? If so, were the sumps properly fenced? Were they properly screened?

The attention to detail that marked Pompa's inspection might have had some roots in her background before going to work in the oil industry. A graduate of Western High School in Anaheim in 1978, Pompa had worked for almost three years as a house painter and cashier, meanwhile continuing her education as a part-time student at Orange Coast College in Costa Mesa. In May 1981, she had joined the U.S. Army, spending the following four years in communications, including more than two years in Germany. When she left active duty in 1985 with the rank of Sergeant, she had joined the California National Guard.

For Syndi Pompa, a former roustabout with an oil company, there was nothing new about pumping units, though now she was the one pointing out deficiencies rather than remedying them. (William Rintoul.)

Completing the lease inspection in the Poso Creek field, Pompa headed back to Bakersfield, where her schedule called for the inspection of three water-disposal wells on the grounds of a refinery in the Fruitvale field immediately west of town and a pair of disposal wells in another plant. Though her visits to both facilities were unannounced, she found the well sites clean.

Back in the office, she would finalize the notes she had taken, make a report on the leases she had visited and note any deficiencies. The engineering staff would take it from there, sending off notices of any problems to operators and giving them a time limit in which to remedy the situation.

During the week that Kathy Roush, Joe Perrick, Dwight Isenhower, Bill Winkler and Syndi Pompa were performing their assignments, the Division of Oil and Gas received notices from various operators, including major companies and independents, to drill 53 new oil and gas wells. In the six Division oil and gas offices throughout the state, personnel processed the notices to ensure compliance with state regulations.

Of the notices, nine were for exploratory wells, often called wildcats. The wildcats would be drilled to look for new pools in existing fields, extensions of fields or for new fields. Six were to be drilled in the Sacramento Valley gas province as tests for new production northward from Stockton to the Orland area near Corning. The other wildcats were to be drilled near Camp Roberts in the Salinas Valley and in the Merced area in the northern end of the San Joaquin Valley and the Wheeler Ridge area at the southern end of the valley.

The remaining 44 wells were classified as development wells. One was for the Wilmington field, with a proposed drill site six miles southeast of the discovery well that had opened the Wilmington field in 1932, proving up California's largest oil field.

Other development wells were in heavy crude fields in the southern San Joaquin Valley, including the Kern Front, Kern River, Midway-Sunset, Mount Poso and South Belridge fields. The Kern River wells were to be drilled on a lease four miles northwest of the hand-dug well that opened the billion-barrel field in the spring of 1899. The Midway-Sunset

Were there leaks in the gathering system? Oil on the ground? It was Syndi Pompa's assignment to find out. (William Rintoul.)

wells were to be drilled on a property 14 miles west of the shoreline of Buena Vista Lake where the Smithsonian Institution, in the depths of the Great Depression, had directed the dig that turned back the pages of time for a better understanding and appreciation of those who had been among the first to make use of California's oil resource.

Since the development wells were to be drilled in areas of proved production, virtually all were assured of successful completion. Given the longevity of existing wells in the fields, it seemed certain many of the wells would be making a contribution to California's oil production on into the next century, with the probability some would be producing when the Division of Oil and Gas marks its centennial.

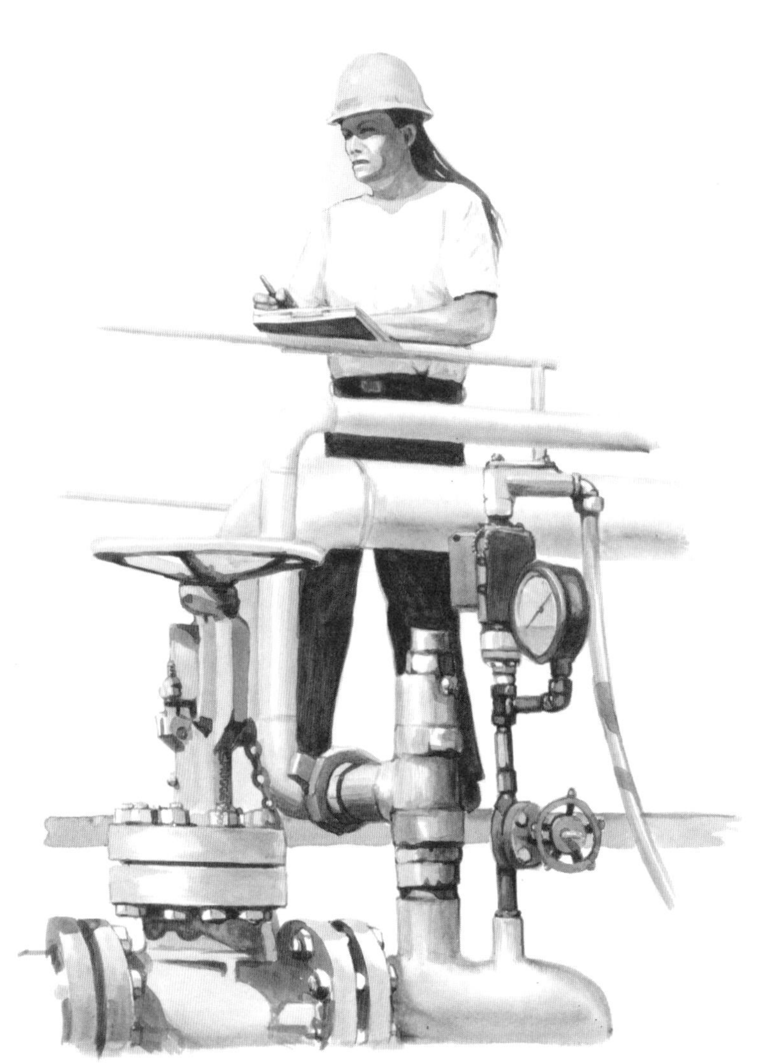

Division of Oil and Gas Managers

STATE OIL AND GAS SUPERVISORS

	Name	Years
1.	R. P. McLaughlin	1915 to 1920
2.	R. E. Collom	1921 to 1923
3.	R. D. Bush	1923 to 1954
4.	E. H. Musser	1954 to 1962
5.	E. R. Murray-Aaron	1962 to 1968
6.	F. E. Kasline	1968 to 1971
7.	J. F. Matthews, Jr.	1971 to 1975
8.	H. W. Bertholf	1975 to 1976
9.	M. G. Mefferd	1976 to Present

CHIEF DEPUTIES

	Name	Years
1.	R. E. Collom	1917 to 1919
2.	R. D. Bush	1919 to 1920
3.	J. B. Case	1920 to 1940
4.	E. Huguenin	1940 to 1950
5.	H. V. Dodd	1950 to 1953
6.	E. H. Musser	1953 to 1954
7.	E. J. Kaplow	1954 to 1955
8.	E. R. Murray-Aaron	1955 to 1961
9.	F. E. Kasline	1962 to 1968
10.	W. C. Bailey	1968 to 1969
11.	J. F. Matthews, Jr.	1970 to 1971
12.	F. E. Kasline	1971 to 1974
13.	J. R. Weddle	1974 to 1975
14.	M. G. Mefferd	1975 to 1976
15.	S. Cordova	1977 to 1987
16.	K. P. Henderson	1988 to Present

DEPUTY SUPERVISORS

*District 1 (Long Beach)**

	Name	Years
1.	R. B. Moran	1915 to 1917
2.	M. J. Kirwan	1917 to 1919
3.	I. V. Augur	1919 to 1921
4.	M. H. Soyster	1921 to 1923
5.	C. C. Thoms	1923 to 1925
6.	E. Huguenin	1925 to 1941
7.	E. H. Musser	1941 to 1954
8.	R. W. Walling	1954 to 1956
9.	W. C. Bailey	1957 to 1968
10.	J. F. Matthews, Jr.	1968 to 1969
11.	M. B. Albright, Jr.	1970 to 1971
12.	W. L. Ingram	1971 to 1975
13.	M. B. Albright, Jr.	1975 to 1979
14.	R. A. Ybarra	1979 to 1981
15.	J. L. Hardoin	1982 to 1983
16.	V. F. Gaede	1983 to 1987
17.	R. K. Baker	1988 to Present

*The district office was located in Los Angeles from 1915 to 1957. In October 1957, the office was moved to Inglewood. In November 1972, the office was moved to Long Beach.

*District 2 (Ventura)**

	Name	Years
1.	I. V. Augur	1917 to 1919
2.	R. N. Ferguson	1919
3.	L. Vander Leck	1919 to 1920
4.	C. C. Thoms	1920
5.	H. B. Thomson	1920 to 1923
6.	E. Huguenin	1923 to 1924
7.	H. A. Godde	1924 to 1925
8.	C. C. Thoms	1925 to 1946
9.	E. J. Kaplow	1946 to 1954
10.	S. H. Rook	1954 to 1962
11.	D. E. Ritzius	1962 to 1975
12.	J. L. Hardoin	1976 to 1981
13.	M. W. Dosch	1982 to 1987
14.	P. J. Kinnear	1987 to Present

*District activities were handled by District 1 from August 1915 to November 1917. The office was moved from Santa Paula to Ventura in September 1985.

*District 3 (Santa Maria)**

1.	R. E. Collom	1915 to 1917
2.	H. W. Bell	1917 to 1920
3.	M. H. Soyster	1920 to 1921
4.	F. D. Gore	1921 to 1924
5.	R. E. McCabe	1924 to 1929
6.	S. G. Dolman	1930 to 1952
7.	W. C. Bailey	1952 to 1957
8.	C. L. Barton	1957 to 1969
9.	M. B. Albright, Jr.	1969 to 1970
10.	A. G. Hluza	1970 to 1971
11.	J. L. Zulberti	1971 to 1983
12.	K. P. Henderson	1983 to 1988
13.	H. P. Bopp	1988 to 1990

*The district office was located in Santa Barbara from 1930 to 1942.

*District 4 (Bakersfield)**

1.	C. Naramore	1915 to 1916
2.	R. N. Ferguson	1916 to 1919
3.	L. Vander Leck	1919
4.	T. D. Kirwan	1919 to 1920
5.	C. C. Thoms	1920 to 1923
6.	W. W. Copp	1923 to 1924
7.	E. Huguenin	1924 to 1925
8.	H. A. Godde	1925 to 1929
9.	E. H. Musser	1929 to 1941
10.	H. V. Dodd	1941 to 1950
11.	R. W. Walling	1950 to 1953
12.	G. G. Peirce	1953 to 1969
13.	G. W. Hunter	1969 to 1971
14.	A. G. Hluza	1971 to 1975
15.	G. W. Hunter	1975 to 1981
16.	A. G. Hluza	1981 to 1985
17.	E. A. Welge	1986 to 1990
18.	H. P. Bopp	1990 to Present

*The first district office was located in Taft. The Bakersfield office was opened in October 1916. The Taft office was closed in June 1971, and operations were consolidated in Bakersfield.

District 5 (Coalinga)

1.	M. J. Kirwan	1915 to 1917
2.	R. D. Bush	1917 to 1919
3.	J. B. Case	1919 to 1920
4.	R. M. Barnes	1920 to 1924
5.	V. H. Wilhelm	1924 to 1926
6.	R. L. Keyes	1926 to 1928
7.	E. H. Musser	1928 to 1929
8.	H. V. Dodd	1929 to 1941
9.	E. J. Kaplow	1941 to 1946
10.	R. W. Walling	1946 to 1949
11.	G. G. Peirce	1950 to 1953
12.	E. R. Murray-Aaron	1954 to 1955
13.	C. W. Corwin	1955 to 1974
14.	F. L. Hill	1975 to 1980
15.	R. F. Curtin	1981 to Present

*District 6 (Woodland)**

1.	F. E. Kasline	1957 to 1962
2.	R. M. Barger	1962 to 1975
3.	J. C. Sullivan	1975 to 1988
4.	R. A. Reid	1989 to Present

*The district office was established in 1957. Previously, the region was included in District 5.

GEOTHERMAL OFFICERS

1.	D. N. Anderson	1970 to 1976
2.	A. D. Stockton	1976 to 1985
3.	R. P. Thomas	1986 to Present

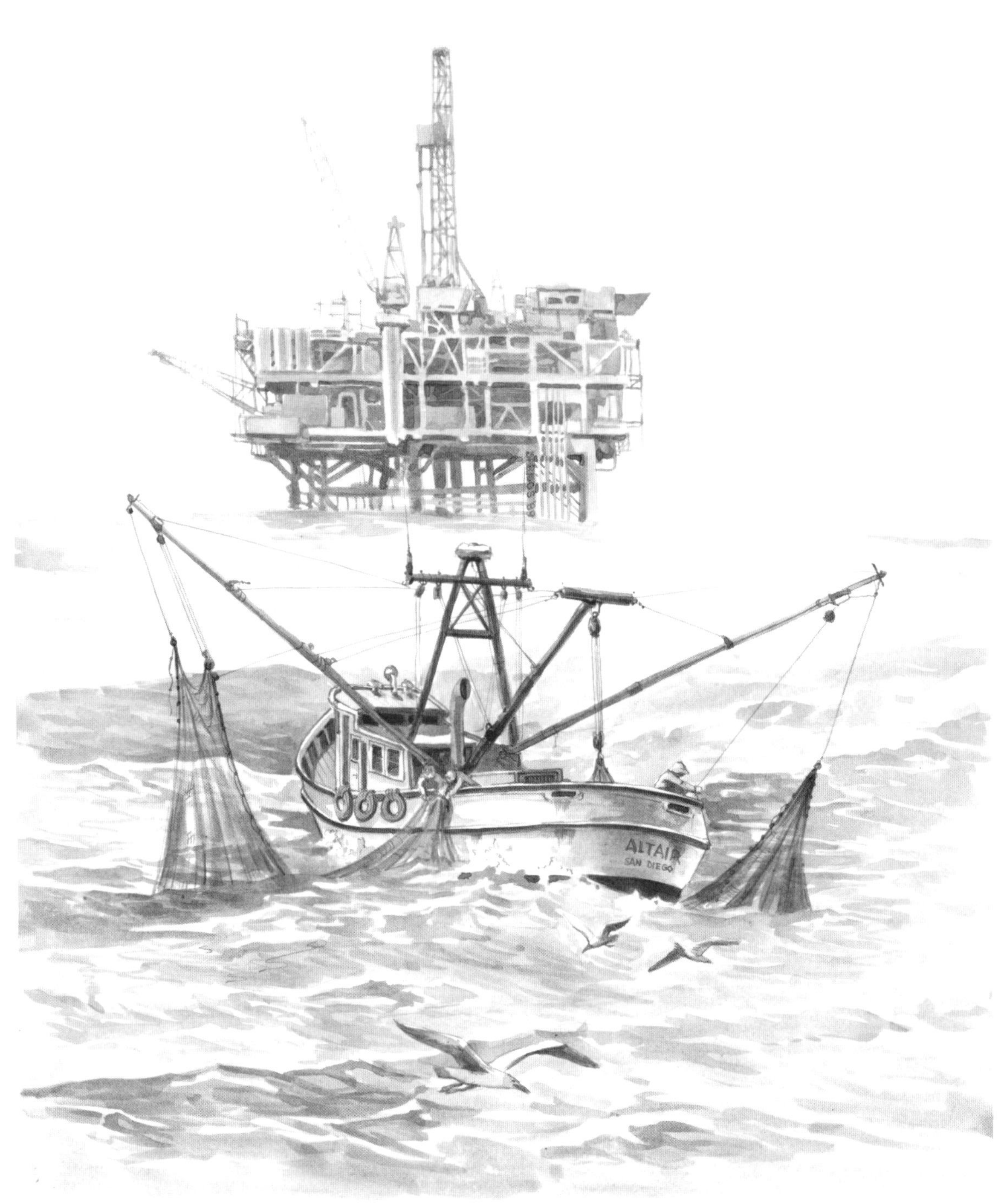
ALTAIR
SAN DIEGO

Acknowledgments

We wish to thank the following people for their interest in and contributions to this project: Burt Amundson, Esther Baisden, Margaret D. Binkley, Timothy S. Boardman, Hal P. Bopp, William E. Brannon, Kenneth M. Carlson, Frances Clark, Simon Cordova, Michael F. Cummings, Richard F. Curtin, Fred Deaubler, Ray Dickinson, Ronald DiPippo, Krug Dunbar, Eleanor Erikson, Nancy Gaede, Michael Glinzak, William F. Guerard, Jr., Robert S. Habel, John O. Harries, Kenneth P. Henderson, Thomas C. Hinrichs, Regina Hulsy, George Hunter, Kenneth Hunter, Bruce W. Manley, M. G. (Marty) Mefferd, Stephen P. Mulqueen, Averill Munger, Dennis Olmsted, Betty J. Parker, Robert A. Reid, Janet Southwick, A. Jean Sporer, Kenneth F. Stelling, A. D. Stockton, John C. Sullivan, Georgia Thomas, Richard P. Thomas, Gael Troughton, James Vantine, and Mary C. Woods.

Further Reading

Beaton, Kendall, *Enterprise in Oil: A History of Shell in the United States*, Appleton-Century-Crofts, Inc., 1957.

Division of Oil and Gas, *California Oil, Gas, and Geothermal Resources, An Introduction*, 1983.

Division of Oil and Gas, *Publications of the California Division of Oil and Gas*, 1990.

Hodgson, Susan F., *Geothermal in California*, Division of Oil and Gas, 1988.

Hodgson, Susan F., *Oil and Gas in California*, Division of Oil and Gas, 1985.

Hutchinson, William H., *Oil, Land and Politics: The California Career of Thomas Robert Bard*, University of Oklahoma Press, 1965.

Jones, Charles S., *From the Rio Grande to the Arctic: The Story of the Richfield Oil Corporation*, University of Oklahoma Press, 1972.

Latta, Frank, *Black Gold in the San Joaquin*, Caxton Press, 1949.

Rintoul, William, *Drilling Ahead: Tapping California's Richest Oil Fields*, Western Tanager Press, 1981.

Rintoul, William, *Oildorado: Boom Times on the West Side*, Valley Publishers, 1978.

Rintoul, William, *Spudding In: Recollections of Pioneer Days in the California Oil Fields*, California Historical Society, 1976.

Tompkins, Walker A., *Little Giant of Signal Hill*, Prentice Hall Inc., 1964.

Waddell, Paul R., and Robert F. Niven, *Sign of the 76*, Union Oil Company of California, 1976.

Welty, Earl M., and Frank J. Taylor, *The 76 Bonanza: The Fabulous Life and Times of the Union Oil Company of California*, Lane Magazine & Book Company, 1966.

White, Gerald T., *Formative Years in the Far West: A History of Standard Oil Company of California and Predecessors Through 1919*, Appleton-Century-Crofts, 1962.

White, Gerald T., *Scientists in Conflict: The Beginnings of the Oil Industry in California*, The Huntington Library, 1968.

Index

abandonment of wells 154-155
accident rate 53
Adams, Ansel 55
Adams Canyon 6,13
Agwiworld 73
Aladdin Oil Company 12
Alamitos Land Company 44,46
Alamitos No. 1 44-45
Alaska Pioneer Oil Company 31
Alegria Offshore oil field 143
Alexander, Ford 37-39
Alexander No. 1 50
Allied Chemical Corporation 93
Altamont Pass 81
Amaurot Oil Company 29
Amerada Petroleum Corporation 65,131
American Association of Petroleum Geologists 106
American Eagle Oil Company 12
American Oilfields Company Ltd. 31
American Petroleum Company 31
American Petroleum Institute 49,72,79
Anadarko Basin 87
Anchor Oil Company 119
Apollo Oil Company 12,29
Army Air Corps 71
Army-Navy Petroleum Board 79
Arnold, Ralph 32-33
Arthur Hotel 33
asphalt 3
Asphalto 1,3-4,10
Assembly Bill 2209 145
Associated Oil Company 22, 34-35,49
Associated Press 55
Astronaut Islands 95
Atchison, Topeka and Santa Fe Railway Company 31
Atlantic Oil Company 85
Audubon Society 145,147
Austin, Dr. Carl 131-132
Automobile Club of Southern California 54
Axelson Manufacturing Company 74

Babbitt No. 1 101,103
Baird, Lieutenant Colonel A. R. 74
Bakersfield 10-11,36-37,62,65,79, 137
Bakersfield Californian 158
Bakersfield Morning Echo 37
Baldwin No. 4 31
Bankline Oil Company 74
Bard, Thomas R. 5
Barker, John 10
Barnsdall Oil Company 54,99
Bell, Robert S. 141
Belmont Offshore oil field 105
Belridge Oil Company 24,120
Belridge oil field 78
Benny, Jack 139
Berry No. 1 64
Beverly Hills oil field 42,138,141
Big Bend 133
Big Four Oil and Gas Company 43
Big Sulphur Canyon 123
Binkley, Bill 71,77
Binkley, Margaret 71,76-79
Binkley, Mary 71
Black Tuesday 55,57
blowout prevention and control 36-39,49-53,151-153
Blue Bird Oil Company 12
Blue Goose gusher 10
Blunt-Nosed Leopard Lizard 148
bonding for wells 62
Boyd, W. R. Jr. 72
Boyle Heights oil field 139
Bradley 81
Brawley geothermal field 131
Brea 146
Brea-Olinda oil field 22,31,42
Bridge, A. F. 49
Brown Drilling Company 104,127
Bryan, William Jennings 29,123
Buchanan's Pavilion 59
Buena Vista Hills 1,77
Buena Vista Lake 1,163
Buena Vista oil field 146-148
Buena Vista Petroleum Company 1
Burbank, Luther 123
Burge, Henry C. 139
Bush, President George 84
Bush, R. D. 26,49,86-87,165-166
Butler-Richardson 118
Byrom, Stan 159

cable tools 4,22
Cal Canal oil field 62
Caliente Offshore gas field 143
California Energy Company 132
California Environmental Quality Act 133-134
California Offshore Oil and Gas Seeps 144
California Oilfields Ltd. 20-21
California Oil World 11,15,23, 25,54,59,64,74,76,78,82,120,135
California Operators General Committee on Gas Conservation 49
California Research Corporation 112-113,115
California, State of 92-93
California Star Oil Works Company 5,84
Canadian Pacific 23
Canal oil field 67
Canfield, Charles A. 7
Canfield Ranch oil field 67
carbon dioxide 130
Carpinteria Offshore oil field 143
Carter, Colonel John J. 17
Carter Oil Company 17
Casa Diablo geothermal field 133
Casmalia oil field 120
Cat Canyon oil field 120
C. C. Harris Oil Company 23
Cedarville 133
Central Valley Regional Water Quality Control Board 145
Chalk Mountain 80,83
Chanslor-Canfield Midway Oil Company 14,30-31
Chanslor-Western Oil & Development Company 117
Cheviot Hills oil field 138
Chevron Corporation 5
Chevron USA Inc. 84,120,131,146
China Lake Naval Weapons Center 131
Chumash 1
Civil War 5,17
Clark, E. W. 49
Clyde Hall Drilling Company 86
Coal Oil Point Offshore oil field 143
Coalinga 6,20-21,29,54,61,74, 119-120
Coalinga oil field 10,14,18-21,68, 115-116
Coalinga Water Arbitration Association 21,23-24
Collom, Roy E. 25,42,47,165-166
Commander Heim Bridge 90

Committee on Manufacturing, Oil and Mining Industry 91
Conception Offshore oil field 143
Confusion Hill 85-86
Conoco Inc. 147
Conservation Committee of California Oil Producers 72,89
Continental Oil Company 67,82, 105,112
Coso geothermal field 131-133
Coso Hot Springs 131
Coyote Hills oil field 23,31
Crestmont Oil Company 116
Crites, Angus 10
Crosby, Bing 139
Cuarta Offshore oil field 143
Cunningham-Shell Tidelands Act of 1955 106-108
Curzon, Lord 41
CUSS I 105-106
Cuyama Valley 80,82-84,97
Cymric oil field 120

Daily Californian 10
Daily Midway Driller 29
D & D Oil Company 118
Davies, Ralph K. 71-72,79
Davis, Dr. E. Fred 67
deepest wells 58-59,63-64,67-68, 78-79,87-88,151
Denverton gas field 88
Department of Fish and Game 145, 147
Department of Petroleum and Gas *(See also Division of Oil and Gas)* vii-ix,24-32,34-35,39,42, 47,53-54
monthly and annual reports: 27,33,41-43,47
personnel and organization: 25-27,29-30,34,36,44,47,165-166
Desert Hot Springs 133
Deukmejian, Governor George vii
Devils Den Area 29
disposal of oilfield waters *(See also injection)* 156-160
Division of Oil and Gas *(See also Department of Petroleum and Gas)* vii-ix,53,60-63,71,81,86, 97,103,117,132,134,144, 151-163
construction-site review and well-reabandonment program: 147
environmental emphasis: 134, 144-148
geothermal authority: 128
monthly and annual reports: 63,95,97-99,107-108,120,128
Offshore Unit: 144-145
personnel and organization: 71,76-79,87-89,132-133, 150,165-166
Doheny, Edward L. 6-10,31,33-34, 141
Doheny Pacific Petroleum Company 31
Dollar Lines 63
Dominguez oil field 47,61
Dos Cuadras Offshore oil field 144
Drake, Colonel 4
Dresser Industries 84
Driscoll, Denny 20
Dunlap, Bob 113
Dunlap, F. P. 15
Dunnigan Hills gas field 88
Durham gas field 88
Dye, Gary 154

E. A. Clampitt Oil Company 118
Earth Energy 129
East Coalinga Extension oil field 61,68
East Coyote oil field 42,147
East Mesa geothermal field 131,133
East Texas oil field 90,93
East Whittier 23
East Wilmington 94
Eckis, Rollin 83
Eddy, Nelson 138
Edison oil field 54,119-120, 157-158
Elk City, Oklahoma 87
Elk Hills oil field 35-39,78
Elliott No. 1 54
Elliott, William B. 123
Elwood, James Munroe 11
Elwood, Jonathan 11
Elwood oil field 54,61,73-74, 99-100,108
Environmental Defense Fund 147
Esso Research & Engineering Company 112
Eureka 105

Fairfield Associated Oil Company 23
Federal Emergency Relief Administration 1
Fellows 14,21
Folsom, David M. 29
Ford Motor Company 58
Ford, President Gerald 129
Fort Bidwell Indian Reservation 133
Four Oil Company 29,153
Fox, Charles P. 11
Freeport gas field 88
Fremont, John C. 123
Fried, Julius 14
Fuqua, W. 52

Galt 28
Gas Act 53,60-62,87
gas companies: *See names of individual companies*
gas fields: *See names of individual fields*
gas injection 61
gas wastage 47,49,53,60-62,87
Gaviota Offshore gas field 143
Gene Reid Drilling Inc. 136-137
General Petroleum Corporation 23, 31,33,52,54,63-64,111-113, 115,120
GEO East Mesa Limited Partnership 131
George F. Getty Inc. 51-52
geothermal companies: *See names of individual companies*
geothermal fields: *See names of individual fields*
Geothermal Hot Line 132
Geothermal Science and Technology, A National Program 132
Getty Oil Company 62,147
Geysers geothermal field 122-130,132-134
Geysers Power Plant Unit 1 126
Giant Kangaroo Rat 148
Gilmore Island 136
Glendale 10
Gollehon, Ernie 116
Gordon Drilling Company 85
Gossett, Ray 86
Grant, John D. 123-125,127
Grant, President Ulysses S. 123
Great Depression 1,56-60,63-65, 100-101,163
Greeley oil field 66,157
Green Cabins area 146
Greenville 133
Grew, Priscilla 62
Guadalupe oil field 120
Gulf Oil Corporation 112,144

Hadley, George 83

Halliburton steam generator 116
Hambleton, Jim 1
Hamilton, Fletcher M. 21,23-24,27
Hammer, Dr. Armand 141
Handbook of Yokuts Indians 2
Hanson, Judge Clarence M. 86-87
Harbor Drilling Company 63
Hartley, Fred L. 129
Hartnell No. 1 14
Hathaway No. 7 58
Hawkeye State Oil Company 12
Hay No. 4 36
Hay No. 7 36-39
Hayes, Frank 44
Hayward & Coleman 3-4
Hazardous and idle-deserted wells 145-147
H. B. No. 15 103
Heber geothermal field 130-131, 133
Hendershott, Clarence 37
Henderson, A. D., Jr. 59
Hillcrest Country Club 138
Hillman, Fred 23
Hodgson, Susan 77
Homan A No. 81-35 84
Homan, Glenn 84
Home Oil Company 10
Honolulu Consolidated Oil Company 31
Honolulu Oil Corporation 112
Hoover, President 58
Hope, Bob 139
Horne 52
Howard, Oscar 58
Hudson, Joe B. 106
Humble Oil & Refining Company 93,107-108
Humboldt County 4-5
Huntington A No. 1 42
Huntington Beach 42,146
Huntington Beach Land Company 42
Huntington Beach oil field 42-43,79,101-103,107,146
Huntington National Oil Company 43
Huntington Sure Shot Oil Company 43

Ickes, Harold L. 71-72
Imperial County 128-131,133
Independent Oil Producers Agency 23
Indian & Colonial Development Company 19,23
Indian Petroleum Corporation 100
Indian Valley Hospital 133
Indiana 13
Industrial Accident Commission 53
Inglewood oil field 47,147
injection of oilfield waters *(See also disposal)* 156-160
International No. 7 34
Interstate Oil Company 31
Isenhower, Dwight 155-158,162
Island Alpha 94
Island Bravo 94
Island Chaffee 95
Island Charlie 94
Island Delta 94
Island Freeman 95
Island Grissom 95
Island White 95

Japan 72,83
Japanese 71
submarines: 73-74
Jergins Oil Company 82
Jewett and Blodget 10
J. J. Elmore Geothermal Power Plant 130
Johnson, H. R. 32
Johnston formation tester 64
Jones, Charles S. 83,90
Jones, George A. 147
Jones-White Act 58
Juanita No. 1 85

Kalispel lease 19
Kasline, Fred 121,165-166
KCL-A No. l-29 65,67
KCL-A No. 72-4 87-88
KCL No. 20-13 78
Kettleman Hills oil field 54,60,64
Kettleman North Dome oil field 74
Kernco No. 1-34 66
Kern County 54,147
Kern County Land Company 65-67
Kern County Oil Protective Association 20-21,23-24
Kern Front oil field 120,162
Kern Oil Company 13
Kern River 10-11
Kern River oil field 12-13,16-19, 28-29,76,110-111,115-121, 153-158,162
Kern Trading and Oil Company 23,30
King City 82
Kingsbury, K. R. 57
Kinley, K. T. 37
Klein, R. Julius 57
Klondike Oil Company 10
Knight, Governor Goodwin J. 91
Kuparuk River oil field 121

La Brea tar pits 3
Lacy Trucking Company 104
Lake Elsinore, City of 133
Lakeview No. 1 (Lakeview gusher) 14-15
Lakeview No. 2 Oil Company 23,31
Lakeview Oil Company 14-15
Lanzer, Martin 159
Larderello 123,132
Larry Doheny 73
Las Cienegas oil field 140
Lassen County 133
Latta, Frank 2
Lawndale oil field 54
lease inspection 161-162
Lindsay, Joe 113
Lockwin No. 1 85
Loffland Brothers Company 129,131
Loftus, William 22
Lombardi No. 1 82
London, Jack 123
Lone Star Producing Company 88
Long Beach 89-94,104
Long Beach City Council 93,95
Long Beach Harbor 89,92
Long Beach Harbor Department 89-90,92
Long Beach Naval Shipyard 92
Long Beach oil field *(See also Signal Hill)* 47,49,57,60
Long Beach Women's Club 46
Lord Roberts Oil Company 12
Los Angeles 6-7,36,46,74,84,91-92, 135-136,139,143
Los Angeles Basin 23,42,47, 52,54,57,97,106,111,146
Los Angeles Beautiful 141
Los Angeles City Council 137
Los Angeles City oil field 8-11, 135,141
Los Angeles County 6,23
Los Angeles County Planning Commission 137
Los Angeles Harbor 92
Los Nietos Company 61
Lost Hills oil field 30,78
Lovejoy, Judge 1
Lucas, D. D. 118
Lynch Canyon oil field 120

Lyons Station 5

MacDonald, Jeanette 138
MacMurray, Fred 138
Magma Power Company 125, 128-131
Magma-Thermal Power 126-127
Maidu 2
Mainland 92
Maricopa 10,14,21,112-113
Maricopa Flat 34
Marine Exploration Company 103-104
Martinez refinery 53
Mascot No. 1 57-59,63-64
Mattole River 4
Mattole Valley 1
maximum efficient rate 72
Maxwell Hardware Company, Oil Division 119
Maxwell, W. O. 23
McCabe, B. C. 125-126
McDonald Island gas field 65
McGreghar Land Company 118
McHale, A. G. 85
McKittrick 1,4
McKittrick Oil Company 23
McKittrick oil field 10,23,29
McLaughlin, Roy ix,22-25,27-31, 33-36,41,165
McMillan, Dan 125-126
Means, Thomas 10-11
mechanical integrity test 158-160
Mefferd, M. G. ix,62,147,165
Merchant Marine 58
Midway No. 2-6 14
Midway oil field 6,14,18-20,23
Midway-Sunset oil field 14,18,21, 57,78,111-115,117,119-121,162
Midway Supply 23
Milham Exploration Company 54
Miller, Guy 70
Miocene Oil Company 23,31
Mitchell, P. D. 118
Mobil Oil Corporation 93,113-114, 120,143-144
Modelo Canyon 7
Modoc County 133
Molino Offshore gas field 143
Montebello Oil Company 31
Montebello oil field 31,34,42,61
Monterey County 82
Monterey Island 103-105
Monterey Oil Company 103-104, 106-107
Moody Gulch 6
Morgan, Frank 83
Morgan, J. Pierpont 123
Morris, Wilson 86
Morrison, Nate 118
Morton International 130
Mother Lode miners 3-4
Mount Poso oil field 54,66, 118-120,162
Mountain View oil field 61,119,145
Murphy Oil Company 23
Musser, E. H. 54,57,87,92,165-166

Nacirema well 24
Naples Offshore gas field 143
National 80-B rig 109
National 50-A rig 104
National Steel and Shipbuilding Company 108
Naval Petroleum Reserve No. 1 78
Naval Weapons Center 132
Nelson, DeWitt 61
Nelson-Phillips Oil Company 84-85
Nesa No. 1 49-50
Nevada Petroleum Company 31
New York Stock Exchange 57
Newhall 5-6,141
Newhall refinery 5
Niland 130
Norris, Halvern L. 83
Norris Oil Company 80,83-84
North Coles Levee oil field 67,83
North Kettleman Oil & Gas Company 64
Northern Light Oil Company 12
Notice of Intention to Drill 24

Occidental Petroleum Corporation 62,137,140-141
Oceanic Oil Company 79
O'Conner, John 75
Off, Charles Frederick 14-15
Off, Julia M. 15
Ohio 13
Ohio Oil Company 23,87-88,112
Oil City area 10
oil companies: *See names of individual companies*
oil fields: *See names of individual fields*
oil seeps 1,3,7,144
  artifacts: 1-2
Ojai No. 6 5
Olinda oil field 22
Olympic Corporation 63
Orange County 23
Orcutt 14
Ormat Energy Systems Inc. 131
Otte, Dr. Carel 129
Ottoboni lease 129
Outstanding Oilfield Lease and Facility Maintenance Award 144,147
Owen, Paul 153
Ozark Oil Company 19

Pacific Coast Gasoline Company 118-119
Pacific Driller 106-108
Pacific Gas & Electric Company 126-128,132
Pacific Light and Power Company 8
Pacific States Petroleum Company 31
Packard Bell Electronics Corporation 141
Packard drilling structure 141-142
Pahmit 2
Pala, Andrew 40,45-46
Palmer, Captain C. J. 92
Paloma oil field 87-88,151
Pan American Petroleum Company 31,54,99
Paris Valley oil field 120
Patton, General George S. 84
Pauley Petroleum Inc. 93
Pearl Harbor 71,73
Peavy, Claude E. 67
Peerless Oil Company 12
Peggy Moore No. 10 85
Peirce, Gil 77-78,166
Pennsylvania 4-5,13,17
Pennsylvanian Oil Company 12
People v. *Metcalfe Oil Company* 86
Perrick, Joe 153-156,162
Petroleum Administration for Defense 71-72
Petroleum Administration for War 72,79
Petroleum Development Company 31
Petroleum Securities Company 61
Petrolia 4
Phillips Petroleum Company 143
Pico, Andreas 141
Pico Canyon 5-6,84
Pico No. 4 5,84
Pier J 94
pipelines 13,84
  Coalinga to Martinez: 54
  Kettleman North Dome to San Francisco: 74

Piru 7
Pitts, E. A. 86
Placerita Canyon 84
Placerita oil field 81,84-86,88
Platform Hazel 107-109
Playa del Rey oil field 48,61,74-75
Pleasant Creek gas field 88
Plumas County 133
Point Conception 108
Point Richmond 13
Pollok, A. J. 23
Pompa, Syndi 160-162
Poso Creek oil field 162
Price, Andrew 122,124-125
Prince Edward Oil Company 12
Prince of Wales 123
Producers Oil Corporation of America 118-119
Producing Properties Inc. 118-119
Production Committee 72
Project HORSE 117
Prosperity Oil Company 12
Providential Oil Company 31
Prudhoe Bay oil field 121
Puente Hills 6
Pure Oil Company 128-130
Pyles, E. E. 72

Rancho Ojai 5
Rancho Park 139
Ranger Petroleum Corporation 63
Ranger, Texas 38
Recovery Oil Company 31
Red Star Petroleum Company 31
reflection seismograph 65-66
Regional Water Quality Control Board 158
Reid, L. H. 67
Reward 23
Rhoten, Duwain 117
Richardson and Bass 87
Richfield Oil Corporation 73,80, 83-84,90,93,105,108,143
Richfield oil field 42
Rincon 104-105
Rincon Island 108
Rincon oil field 54,99 101,108
Rincon "steel island" 100-101
Rio Bravo oil field 66,68
Rio Grande Oil Company 99
Rio Viejo oil field 151
Rio Vista gas field 65,74,88
River Island gas field 88
Riverside County 133
Riverside Portland Cement Company 31
Rockefeller, John D. 17
Rohl-Connolly Company 100
Rolph, Governor James 103
Romany oil 22
Roneck 52
Roosevelt, President Franklin D. 71, 73-74
Roosevelt, President Theodore 123
Rotary Club of Long Beach 90
rotary drilling 81
Round Mountain oil field 119
Roush, Kathy 151-153,162
Royal Dutch Shell 71
Rubbert, C. E. 118
Rubel, Cy 137
Russell A No. 28-5 84
Russell Ranch oil field 83-84

Sacramento Oil Company x
Sacramento Valley 88,162
Salinas River 81
Salinas Valley 81-82
Salt Lake oil field 136
Saltmarsh Canyon 4
Salton Sea geothermal field 130,133
Salton Sea Unit 3 130
San Ardo 81
San Ardo oil field 82,97,120
San Bernardino County 133
San Clemente Oil Company 54
San Fernando Mission 141
San Francisco 3-6,12-13,21
San Joaquin Kit Fox 134,148
San Joaquin Oil Company 13
San Joaquin Valley 1,6,10,23,80, 82-83,87-88,97,117-118,121, 137,145
Sansinena oil field 136-137
Santa Barbara 96-97,99
Santa Barbara Channel 97,100,108
Santa Barbara Channel blowout 144
Santa Barbara County 23,33,54,99
Santa Cruz Mountains 6
Santa Fe 23
Santa Fe Energy Company 117
Santa Fe Springs oil field 42, 49-51,57-58,60
Santa Maria 14
Santa Maria Basin 106
Santa Maria Crude Oil Company 14
Santa Maria oil field 18,28,147
Santa Maria Oilfields, Ltd. 31
Santa Paula 3,6
Santa Paula oil field 32
Sawtelle oil field 137
Schwennesen, Alvin 44
Scofield, Demetrius G. 5
Scott, Arthur 83
Seaboard Oil Company 107
Seal Beach 103-104
Seal Beach oil field 47,103,147
Sespe oil field 146
SFS No. 17 51
Shaffer equipment 56
Shaffer-Warner equipment 59
Shamrock gusher 10
Shasta County 133
Shell, Assemblyman Joseph C. 106
Shell Development Company 112
Shell Oil Company 21,30,40,43-47, 49-50,53-54,57,65-67,82,93,105, 111,115-119,135-137,144
Siegfried, H. N. 131
Signal Hill *(See also Long Beach oil field)* 40,44-50,53,62,104
Signal Oil & Gas Company 138-140
Silver Tip No. 1 14
Simms, Pete 112
Sisquoc 33
Sisquoc River 33
Slocum & Company 32
Slocum, Thomas A. 32
Smith, Andy 125
Smithsonian Institution 1,163
Somavia, Ramon 85
South Belridge oil field 64,78, 111-112,115,120-121,162
South Coles Levee oil field 67,78
South Cuyama oil field 84
South Elwood Offshore oil field 143
Southern California Edison Company 131
Southern California Gas Company 74,146
Southern Counties Gas Company 49
Southern Cross Oil Company 12
Southern Pacific Company 30
Southern Pacific Railroad 5,12,98
Southern Sierras Power Company 131
Sovereign Oil Company 12
Spacing Act 62,86
Spacing among wells 62
Spears, B. J. 86
Spud Inn Restaurant 45
Standard-Humble Summerland State No. 1 109
Standard Oil Company 13,17-18, 20-21,23,25,29-30,33,35-39,

42-43,57-59,63-66,70-71,75,78, 81,93,104-109,112-113,116, 141-143
Stanford Brothers 3-5
Stanford University 41
State Indian Museum 2
State Lands Act of 1938 97,103
State Lands Commission 103, 106-108
State Mineralogist 24,27
State Mining Bureau 21-22, 24-25,97
State Oil and Gas Supervisor 24,27,43,53,89,92,145
Steinbeck, John 81
Steinhilber No. 2 52
Stevens A No. 1 65-67
Strand oil field 67
Stuck, John 25
Stutz Motor Company 58
Submarex 105
Subsidence 89-93,95,97
Subsidence Control Act of 1958 91-92
Suiter, Gordon P. 13
Sullivan, J. 52
Sulphur Mountain 3
Summerland 97
Summerland Offshore oil field 106-109,143
Summerland oil field 96-99, 107,146
Summers, Mrs. Emma 8
sumps 145-146
Sun Drilling Company 138
Sunray Mid-Continent Oil Company 112
Sunset oil field 6,13
Superior Oil Company 52,101,105

Taft 1,20-22,29,34,37-38,57,59-60, 71,78
Taft office, Dept. of Petroleum and Gas 25,28-29,34,71,78-79
Taft, President 35
Taylor, Frederick 5
Taylor, Reese 72
Temblor Range 1
Ten Section oil field 65-67
Terminal Island 92,95
Terminal No. 1 63
Texas 13
Texas Company 82,104,112,143
Texaco Inc. 82,93,144
Thacher, John H. Jr. 113
Thelen, Max 29
Thermal Power Company 125-126, 128-129
Thornburg, Dwight 44
THUMS Long Beach Company 93-95,143
Tide Water Associated Oil Company 76,82
Tidewater Oil Company 112, 115-119
Timken No. 1 131
Titusville, Pennsylvania 4-5
Torrance oil field 47,52,63
Traders Oil Company 19
Trico gas field 75
Trinidad 3
Truitt, Glen 123,125
Tulamni 1
tunnels 3-5
Twain, Mark 123
Twentieth Century Fox 138

Uncle Sam Oil Company 12
Underground Injection Control Program 156-160
Union Mattole Oil Company 4-5
Union Oil Company 13-15,19,31, 33-34,47,49-50,66-68,72,74,82, 93,105,112,128-129,132, 135-137,140,144
Union-Signal-Pacific Electric No. 1 140
unitization 91-92
Universal Consolidated Oil Company 138
Universal Oil Company 31
Unocal Corporation 130-131,147
U. S. Bureau of Mines 53
U. S. Bureau of Reclamation 131
U. S. Congress 35,72
U. S. Fish and Wildlife Service 148
U. S. Forest Service 146
U. S. Geological Survey 35,132
U. S. Navy 35,78,89,131-132

Venice 48
Venice Beach oil field 143
Ventura 3-5,13,44
Ventura Basin 106
Ventura County 6,23,32,54
Ventura Oil Company 118
Ventura oil field 60
Verne Community No. 1 136
Verne, Jules 136
Veterans Administration Center 137
Vidor, King 138
Visalia Midway Oil Company 23
Vonderahe No. 1 130

wages 28,43
Wahumchah 2
War of 1812 74
Wasco oil field 67
water blast 3
waterflood 91-93,95
Watson No. 2 63
Watts, W. L. 6,97-98
Wayne, John 138
well inspections 151-162
West Coyote oil field 42
West Crude Oil Company 118
West Virginia 13
Western Explorer 104-105
Western Interstate Commission for Higher Education 147
Western Offshore Drilling & Exploration Company 109
Western Oil and Gas Association 147
Westlake Park 7
West Shore Company 153
West Side 1, 29,57,78
Wheeler Canyon 4
White, W. T. 88
Whitehall, Ralph R. 118
Whittier 14
Whittier oil field 33
Wild Goose gas field 88
Wiley Canyon 6
Williams, H. L. 97
Willis, Dr. Bailey 41-42
Wilmington oil field 63-64,68, 89-95,97
Wilshire Oil Company 103
Wilson, President 29
Wilson, T. 52
Winkler, Bill 157-160,162
Winters gas field 88
Woodworth No. 1 85
Wootan, Jim 138
Works, Bill 151
World War I 28-30,41,70,73
World War II 68,70-79,81-82,131

Yant, M. R. 85-86
Yokuts 1-2,151
Yorba Linda oil field 111, 115-116,120
Yorty, Mayor Sam 141
Yowlumne oil field 151,153

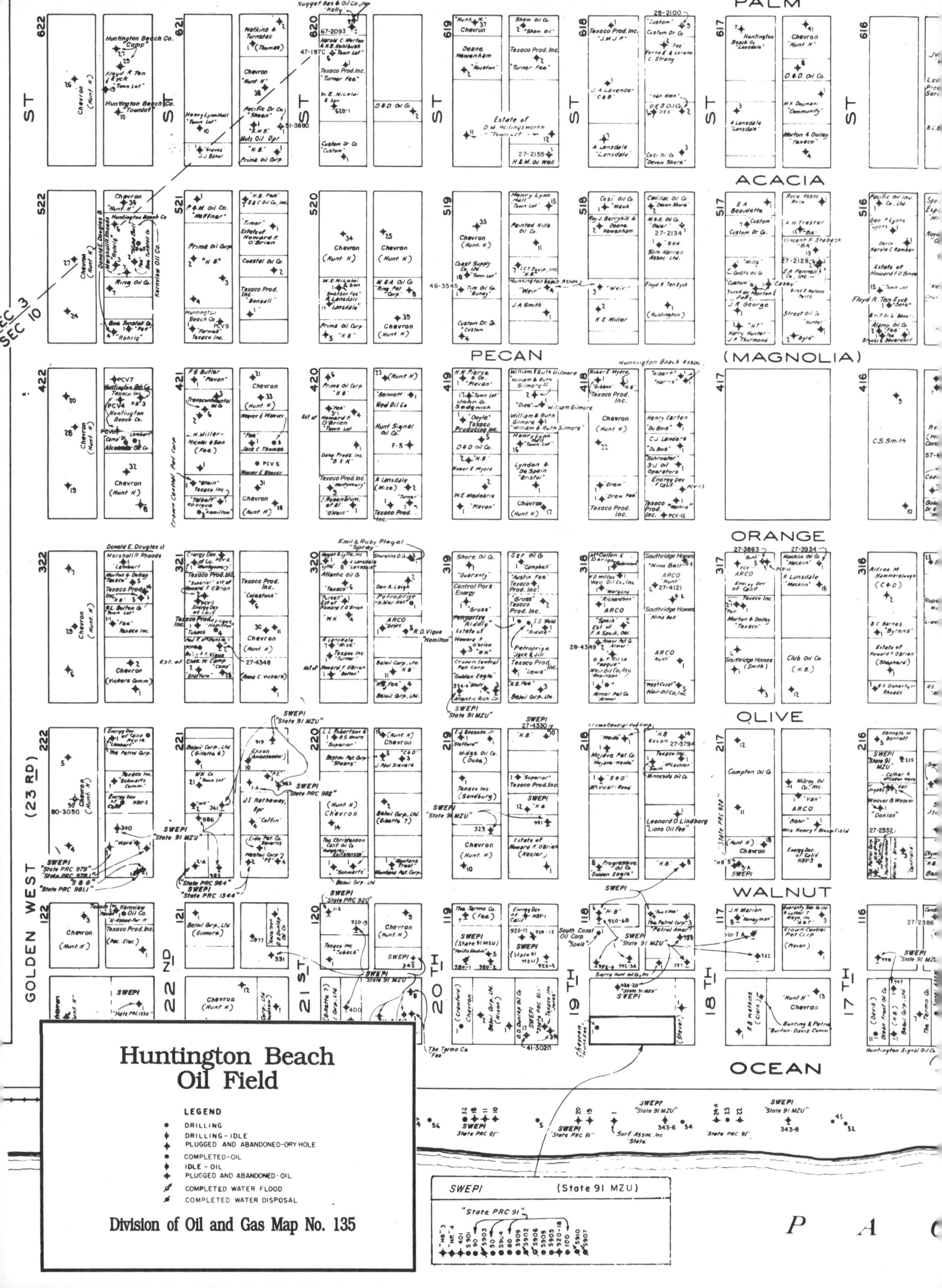
Huntington Beach
Oil Field
LEGEND
DRILLING
DRILLING - IDLE
PLUGGED AND ABANDONED-DRY HOLE
COMPLETED-OIL
IDLE - OIL
PLUGGED AND ABANDONED-OIL
COMPLETED WATER FLOOD
COMPLETED WATER DISPOSAL
Division of Oil and Gas Map No. 135
PALM
ACACIA
PECAN
(MAGNOLIA)
ORANGE
OLIVE
WALNUT
OCEAN
GOLDEN WEST (23 RD)
22 ND
21 ST
20 TH
19 TH
18 TH
17 TH
ST
SEC 3
SEC 10
SWEPI (State 91 MZU)
"State PRC 91"
P A C